WORKING IN AGRICULTURAL MECHANICS

Glen C. Shinn
University of Missouri
Columbia, Missouri

Curtis R. Weston
University of Missouri
Columbia, Missouri

Max L. Amberson
Consulting Editor
Montana State University
Bozeman, Montana

GREGG DIVISION

McGRAW-HILL BOOK COMPANY

New York	Auckland	London	New Delhi	Singapore
St. Louis	Bogotá	Madrid	Panama	Sydney
Dallas	Düsseldorf	Mexico	Paris	Tokyo
San Francisco	Johannesburg	Montreal	São Paulo	Toronto

Library of Congress Cataloging in Publication Data

Shinn, Glen C
  Working in agricultural mechanics.

  (Career preparation for agriculture/agribusiness)
  Includes index.
  1. Agricultural engineering—Vocational guidance.
2. Agricultural machinery—Maintenance and repair—
Vocational guidance. 3. Metal-work—Vocational guidance.
4. Woodwork (Manual training)   I. Weston, Curtis Ross,
1924-   joint author.   II. Title.   III. Series.
S678.4.S53   631.3'023   77-13585
ISBN 0-07-000843-4

620 · 1 S

3 4 5 6 7 8 9 0   VHVH   8 9 8 7 6 5 4

The editors for this book were Susan Horowitz and Susan
Berkowitz, the designer was Sullivan-Keithley, Inc., the art
supervisor was George T. Resch, and the production
supervisor was S. Steven Canaris. It was set in Times
Roman by John C. Meyer & Son. Cover design by Ed Aho.
Printed and bound by Von Hoffman Press, Inc.

CONTENTS

PICTURE CREDITS

Page 6: left, Library of Congress; right, Grant Heilman.
Page 8: top left, International Harvester; other three, Grant Heilman.
Page 9: left, United States Department of Agriculture.
Page 10: both, Grant Heilman.
Page 12: California Canners and Growers.
Page 15: left and top right, International Harvester; bottom right, Grant Heilman.
Page 17: Grant Heilman
Page 18: Grant Heilman
Page 19: United States Department of Agriculture, photograph by Edward S. Trovillion.
Page 20: International Harvester
Page 23: Future Farmers of America, photograph by Susan Berkowitz.
Page 24: Future Farmers of America
Page 26: United States Department of Agriculture
Page 27: International Harvester
Page 40: Keuffel and Esser
Page 117: Grant Heilman
Page 120: Ginger Chih

Front cover: Future Farmers of America
Back cover: Grant Heilman

PREFACE

In the early nineteenth century, most Americans—about 90 percent—lived and worked on farms, producing food, clothing, and shelter for themselves and their families. Today, less than 5 percent of American people live and work on farms and ranches, yet these people feed and clothe the population of the United States as well as provide farm and ranch products for people throughout the world.

The farm family of yesterday worked hard and long hours. It took 35.2 hours to produce an acre of corn. Today, that same acre of corn requires 5.2 hours of labor.

The difference is technology—modern equipment, machinery, tools, and techniques to produce the most efficient yields on our farms and ranches.

But mechanical power alone cannot do the job. People are needed. People operate machines and control mechanical power. People engage in research and develop techniques to continue modernization of agricultural industry.

There is a vast agribusiness complex supporting the production, processing, and marketing of agricultural supplies and products. A recent report by the U.S. Interdepartmental Committee on Employment Opportunities in Agriculture and Agribusiness has stated that only 29 percent of the employment needs are being met by those who are trained. Jobs are available for skilled people.

Working in Agricultural Mechanics: An Instructional Module

Working in Agricultural Mechanics was developed as an instructional module to assist teachers in implementing a meaningful unit in agricultural mechanics as part of a core curriculum. The competencies developed in this module—the understanding, the knowledge, and the technical skills—are the basis of a core curriculum in vocational agricultural education as well as an introduction and essential foundation to advanced programs in production agriculture, agribusiness, and all phases of agricultural mechanics.

The instructional module *Working in Agricultural Mechanics* includes the following components:

- An introductory sound filmstrip that traces the development of agricultural industry and presents its career opportunities
- A student textbook that provides basic skill development in a career context
- A student activity guide that includes application-type questions, tests, and long- and short-term projects (problem-solving situations) to supplement learning of basic concepts
- A set of overhead transparency masters that facilitates classroom discussion and demonstration
- A teacher's manual and key that includes teaching suggestions, answers, and objective tests to assist in implementation of the instructional module

Student Textbook

Working in Agricultural Mechanics gives students exposure through skill development to basic principles of agricultural mechanics as they apply to modern techniques of production agriculture and agribusiness. Through a competency-based approach and behavioral goals, the text permits students to build a foundation of fundamental knowledge essential to the understanding and appreciation of careers in agricultural mechanics and basic skill development in woodworking and metalworking, welding, and tool reconditioning.

The text is divided into units and chapters. An occupational matrix begins each unit. The matrix shows the relationship between jobs in agricultural industry and competencies needed to enter and succeed in those jobs.

On the horizontal axis of the matrix are sample job titles from the seven major career areas of agriculture/agribusiness. The sample jobs are high-frequency jobs that students with vocational agriculture preparation and supervised occupational experience could enter after graduation.

The vertical axis of the matrix shows competencies needed by students who plan to enter the occupations listed. These competencies, which include knowledge, skills, attitudes, and experiences, are designated as *very important, important,* or *not important* for the jobs listed. The competencies listed in the occupational matrices also are developed in the chapters of the text.

Goals stated at the beginning of each chapter help students become aware of learning expectations. End-of-chapter questions are designed to aid students in meeting the stated goals of the chapter. Detailed descriptions using basic vocabulary, step-by-step procedures, and a comprehensive illustration program aid students in building their foundation of skill development. Career examples and information provide relevancy by relating skills to work situations.

Career Preparation for Agriculture/Agribusiness

Working in Agricultural Mechanics is part of an instructional program designed as a core curriculum for students of vocational agriculture. The purpose of the core curriculum is to provide basic principles and entry-level employment skills for agriculture/agribusiness students. The core provides solid foundation for advanced specialization. The titles in the core curriculum are:

- Working in Agricultural Industry
- Working in Plant Science
- Working in Animal Science
- Working in Agricultural Mechanics
- Leadership for Agricultural Industry
- Learning through Experience in Agricultural Industry

A comprehensive program-planning guide has been developed to assist in the implementation of a one- or two-year competency-based core curriculum.

Acknowledgments

Many people assisted in the preparation of this text. However, special recognition should go to Dr. Max L. Amberson, Montana State University, Consulting Editor; Dr. Clinton O. Jacobs, University of Arizona; Dr. Arthur Berkey, Cornell University; and Dr. Richard Linhart, Washington State University, for constructive comments that contributed to the development of this program.

Appreciation also goes to two vocational agriculture teachers whose manuscript reviews helped to strengthen the quality of the material: William Perrow, Appomattox County High School, Appomattox, Virginia, and Terry Gorman, Platteville High School, Platteville, Wisconsin.

The teachability—format, approach, reading level—of the program was confirmed through field testing of selected units.

UNIT I

AN ORIENTATION TO AGRICULTURAL MECHANICS

COMPETENCIES

COMPETENCIES	PRODUCTION AGRICULTURE — Farmer-rancher	AGRICULTURAL SUPPLIES/SERVICES — Produce manager	AGRICULTURAL MECHANICS — Machinery parts clerk	AGRICULTURAL PRODUCTS, PROCESSING, AND MARKETING — Dairy technologist	HORTICULTURE — Florist	FORESTRY — Forest cruiser	RENEWABLE NATURAL RESOURCES — Game farm worker
Describe the role of mechanization in modern agricultural industry	Important	Important	Very Important	Very Important	Important	Important	Important
Identify the six general training areas in agricultural mechanics	Important	Important	Important	Important	Important	Important	Important
List career opportunities in agricultural mechanics within agriculture/agribusiness	Very Important	Very Important	Very Important	Important	Very Important	Very Important	Very Important
Describe the personal characteristics for success in agricultural mechanics	Very Important	Very Important	Very Important	Very Important	Very Important	Very Important	Very Important
Describe the technical knowledge and skills needed to work effectively in agricultural mechanics	Very Important	Important	Very Important	Important	Important	Important	Important

 Very Important   Important   Not Important

Everybody depends on agricultural industry, and agricultural industry depends on machinery, power equipment and tools, and the people who know how to use and care for them.

The people who know how to use and care for machinery, equipment, and tools are engaged in agricultural mechanics. But there is more to agricultural mechanics than that. Agricultural mechanics also means the planning, construction, and maintenance of buildings and other structures; the use of electrical power in agricultural industry; soil and water management; and food processing.

Who needs education in the skills of agricultural mechanics? The farmer, rancher, and farm worker use mechanical skills to keep machines and equipment working properly. They also use agricultural mechanics skills to construct and maintain farm buildings.

Employees in agribusiness also use mechanical skills as part of their jobs. Lawn and garden salespeople should be able to describe the maintenance required for a small lawn tractor, mower, or tiller. Recommending paints and preservatives for outdoor use is also part of the salesperson's job.

Nursery workers are better able to do their jobs if they can properly use hand and power tools. Most employers would like their workers to be able to maintain small gasoline engines.

Forestry workers are another group who make use of mechanical skills. Their jobs often involve work with power saws, hydraulic loaders, trucks, and tractors.

Of course, there are many people employed in the area of agricultural mechanics, and they depend directly on basic and specialized agricultural mechanical skills. The tractor mechanic, machinery mechanic, setup person, and parts clerk are employed in farm equipment dealerships. The conservation contractor and the irrigation service worker use skills in soil conservation and water management. And the farm building construction company employs many agricultural workers who also use mechanical skills.

All these workers depend on the basic skills of agricultural mechanics. In fact, 24 percent of all American workers need some knowledge of agricultural mechanics, according to the U.S. Office of Education.

Ever since 1831, agricultural mechanics has become increasingly important to agricultural industry. That was the year when the McCormick reaper first appeared on the American farm and began to mechanize and revolutionize agriculture. Today, both the production and the processing of food are largely mechanized.

Of course, mechanization has not been accomplished by accident. In other words, it is not true that the machinery was invented first and then agricultural industry tried to find a use for it. The machinery and tools were developed to perform specific tasks; inventors wanted to use machines to replace both human and animal power.

There are two primary reasons for the mechanization of agricultural industry. One has to do with economics and the other with people's natural desire to make life easier for themselves. People in agricultural industry quickly discovered that more food and nonfood items could be produced more economically through the use of machines. So today small truck farmers in New Jersey may use machines to pick tomatoes; cotton farmers in the South have almost all converted to mechanical pickers; and citrus fruit growers in California may use a forklift and box system to harvest their crops. The sight of the giant combines on the Great Plains of the Midwest has become almost symbolic of America's ability to produce enough small grains to feed not only its own people but many of the peoples of the world.

And so it goes. Those in agricultural industry also have used power tools and equipment to do many other related jobs—from repairing machines with welding torches to building a shed with electric saws and drills. The tractor mechanic uses a power wrench; the logger uses a power chain saw or hydraulic shear instead of the old crosscut saw; the nursery worker mixes potting soil with such machines as a front-end loader and power shredder-blender.

As the demand for machinery, power equipment, and tools for work on the farm has increased, so has the demand for more support services. These support services are classified as agribusiness. They include companies that manufacture and distribute fertilizers and pesticides; wholesale and retail stores that sell everything from tractors to nuts and bolts; off-farm shops that specialize in engine maintenance or welding; and contractors who put up buildings, move earth, drill wells, or perform some other special task.

This first unit will examine the importance of agricultural mechanics to agricultural industry and describe the many jobs in which the skills of agricultural mechanics are needed. These are jobs that may be performed on a farm or ranch, in a large plant, laboratory, tool shop, greenhouse, or retail store.

By the end of the unit you should know about the many career opportunities available in agricultural mechanics and what personal qualities and characteristics you will need to succeed in any of them.

CHAPTER 1

OVERVIEW OF AGRICULTURAL MECHANICS

In the physical sciences, *mechanics* refers to the effects and use of energy. The energy could be electricity, which might be used to power giant motors. Then again, the energy could be simple muscle power, and it might be used to manipulate common mechanical drawing tools in creating a blueprint.

As applied to agricultural industry, *agricultural mechanics* refers to the operation and maintenance of machines and equipment used in the industry and the varied mechanical jobs that people perform.

CHAPTER GOALS

In this chapter your goals are:

- To recognize the importance of mechanization and describe how the substitution of machine power for human power has improved the quantity and quality of agricultural products
- To appreciate the mechanical skills required to operate, maintain, and repair agricultural machinery and equipment
- To develop an awareness of the six areas of training that make up agricultural mechanics

- To list several job opportunities available in each of the six training areas

Agricultural Mechanics in Perspective

In the year Cyrus McCormick invented the reaper, 90 percent of the workers on American farms and ranches were performing manual labor. It took 100 persons working long, hard hours to cut 80 acres [32 hectares (ha)] of grain (2000 bushels) with hand scythes. Today, one person operating a combine can harvest the same amount of grain in a single day. (See Figure 1-1.)

By replacing 99 field-workers, the combine did not necessarily force those persons out of agricultural industry. The person who was once needed to cut down grain now may work in a factory making combines, may sell combines to farmers, may service combines, may own or work in a shop that does welding, may work in a grain elevator, or may be employed in any number of other jobs associated with agricultural industry.

Another indication of the tremendous impact machines have had on agricultural industry is found in the following statistics:

Figure 1-1. Machinery replaced much manual labor.

In the period from 1910 to 1914, it took 15.2 hours of labor to plant and harvest 1 acre [0.4 ha] of wheat; from 1970 to 1974, the same cycle took only 2.9 hours. The comparable figures for the production of corn are 35.2 h/acre [88 h/ha] from 1910 to 1914 and 5.2 h/acre [13 h/ha] from 1970 to 1974.

To produce 100 pounds [454 kilograms] of beef during the period from 1910 to 1914, it took 4.6 hours. By 1970, it took only 1.7 hours. The comparable figures for the production of 100 pounds of poultry are 31.4 hours from 1910 to 1914 and 5.2 hours of labor from 1970 to 1974.

During the 1920s, horses and mules provided most of the power in agriculture. Plowing was done by a team of four horses pulling a two-bottom plow that plowed a strip 24 to 28 inches [61 to 71.1 centimeters (cm)] wide. Working from sunrise to sunset, farmers could plow 2 acres [0.8 ha]. Today farmers operate a 150-horsepower tractor that plows a strip 9 feet [3 meters

(m)] wide, and they can plow 60 acres [24 ha] in a 10-hour day.

During the 1920s it was also customary to cultivate a crop at least four times during the growing season. Now, because of weed-killing chemicals applied by machine or airplane, many crops need no cultivating at all.

Not only has mechanization made it more economical and easier to harvest grain, plow fields, and eliminate weeds, but now modern equipment can till the soil, plant, fertilize, and apply insecticides in one operation.

Just as the invention of the combine freed 99 people to do other productive and useful jobs in agricultural industry, so the inventions of the tractor and combination spreaders and the discovery of chemicals have freed people. Because they can plow more acreage with a tractor now, farmers are free to do such things as plan for future development and expansion of farm operations, build a new shed for equipment, plan and install an irrigation system, or perhaps add another enterprise, such as livestock, to their existing operations.

AN ORIENTATION TO AGRICULTURAL MECHANICS

Training Areas in Agricultural Mechanics

Changes in tillage practices, chemical controls, and harvesting are examples of how energy and mechanization have been used to improve the efficiency of production agriculture. But this combination of energy and mechanization has created a need for new skills and services.

Today's worker, for example, must be trained to operate and maintain an expensive and complex combine, tractor, or other equipment. The many workers off the ranch and farm—in agribusiness—also must have specific skills.

A group of engineers and educators has determined what skills and abilities in agricultural mechanics must be learned if men and women are to be prepared to take their places in agricultural industry—in both production agriculture and agribusiness.

This committee has divided agricultural mechanics into six general areas. All these areas require persons who are skilled in performing and supervising tasks that employ energy and mechanics. These six general areas are as follows: (1) agricultural power and machinery, (2) agricultural structures, (3) electrical power and processing, (4) soil and water management, (5) agricultural construction and maintenance, and (6) food processing.

Agricultural Power and Machinery

Agricultural industry uses a wide variety of machinery. Tractors are most common, and there are different kinds of trucks and other vehicles, including those that transport workers. Most of these are powered by large internal-combustion engines that burn either diesel fuel or gasoline.

The industry also uses internal-combustion engines and electric power units for other purposes. A power unit might run a pump, conveyer belt, or an additional generator for power.

Then there are machines that till the soil, fertilize the land, plant the seeds, spray to control insects and disease, harvest what is grown, and process crops for human or animal use. Also there is equipment to grind and mix feeds for animals and to milk cows.

Heavy equipment such as bulldozers, graders, and earth movers is available to construct terraces, waterways, and farm ponds. The farmer or rancher not only conserves soil but also stores water for later use for livestock or crops. In addition, ponds and lakes can be used for recreation and supplemental income if the lake is used for fishing and the adjacent land is used for camping. Heavy equipment also is used to grade the land for irrigation and improved drainage. Lands are graded to attain a uniform slope or grade. This slope may be only 1 in. per 250 ft [2.5 cm per 76.2 m] and not readily visible to the average person, but it provides improved drainage during wet seasons and allows water to run uniformly over the soil to water the crops during dry summers. Land grading has improved crop yields and reduced soil erosion in the United States.

Most farm and ranch workers must be familiar with the uses of energy and mechanical forces. Because of the growth in jobs in agribusiness, many workers must have mechanical skills. They must be able to select different kinds of machines for specific jobs and know how to safely operate, maintain, and repair equipment and different kinds of engines and power units.

Persons employed to operate, maintain, or repair the kinds of machinery and power units described will have to understand the theories and principles that apply to engines, power transmissions, hydraulics, and other drive systems. They may be

Figure 1-2. Workers in agricultural mechanics may be called upon to work on many kinds of equipment.

called upon to service or repair everything from a 3-hp lawn mower to a 300-hp four-wheel-drive tractor equipped with a variety of equipment. (See Figure 1-2.)

Agricultural Structures

Agricultural mechanics also is concerned with the planning, construction, maintenance, and repair of farm buildings.

Some structures in the industry today were not even thought of a few years ago. For example, a few specialty crops such as flowers and vegetables are produced in structures that control the light, temperature, humidity, and other variables that affect production. Many greenhouses have

several acres under glass. The houses have electric heating and cooling systems.

It is not unusual for some animals such as hogs and cattle to be confined inside from their birth to their slaughter, so that they can be protected from extremes in temperature and weather. Research has shown that any agricultural enterprise that raises or keeps animals needs some type of shelter.

Structures also are important to that part of agricultural industry responsible for storing harvested grain. Grain is stored until it can be sold and transported. If it is properly stored, the quality of the grain will be maintained a long time. Elevators that store grain, particularly those in the Midwest, are among the largest in all industries. A Kansas farmers' cooperative owns an elevator that is just 67 ft [20.4 m] short of being one-half mile [0.8 km] long.

Figure 1-3. Construction of agricultural structures is an important part of agricultural industry.

It has a storage capacity of 18 million bushels of grain.

Persons employed to build structures are involved in the planning and construction of new buildings and the maintenance and repair of existing buildings. They may be called upon to install such things as grain elevators, conveyers, water and disposal systems, and feed-handling equipment. Also, they may be expected to recognize grades and standards of building materials and to know what tools to use for different construction jobs.

Electrical Power and Processing

It is hard to say which has had the greater impact on agricultural industry—the internal-combustion engine or electricity. Both have proved indispensable to the expansion of the industry. In recent years, electrical power has become even more important to the industry as the costs of gasoline, diesel fuel, and oil have risen sharply and the supplies have decreased.

In 1930 electricity was available to only 13 percent of rural residents (mostly farmers and ranchers). Rural electrifica-tion was begun in the 1930s by the federal government with the passage of the Rural Electrification Act. This act provided low-interest loans to groups of farmers and ranchers who wanted to organize coopera-tives and build electrical power lines. Rural electrification has progressed to the point that now nearly all rural residents in the nation are served by electricity.

One use for electrical power in ag-ricultural industry is to process agricul-tural products (washing, sorting, packag-ing, etc.). Many of the facilities used to confine livestock indoors use electricity to maintain a carefully controlled en-vironment.

Electricity has made a major differ-ence in dairy farming. Electric lights, for example, have made it possible to milk cows when it is still dark outside. Water is pumped to the milking parlor which is heated by electricity, and the cows' udders are washed with warm water. Then the teat cups are engaged, and the electric-powered milking machines take over. Fresh milk flows into the lines, and then it is transported to a bulk cooling tank. There an electric motor drives the refrigeration unit. Because electricity operates various machines, one or two persons can milk 100 cows in the time it once took to milk 20 cows by hand.

Electrical power is used not only for a variety of purposes in agricultural industry but also by farm or ranch families to operate many conveniences in their homes.

Many persons employed in agricultural industry need to understand electricity and how it is installed and used on the farm and ranch, in the shop to operate power tools and equipment, in the processing plant, and elsewhere.

They may be called upon to install electrical wiring and to maintain and repair electric motors and electric-powered equipment. If they do install electrical wiring, they will have to know how to determine electrical loads on wires, how to select wire sizes, and how to determine the location of electrical power distribution centers. It is also important for them to know the principles and safety regulations governing the use of electricity.

Soil and Water Management

Nearly everyone who is employed in agricultural industry is concerned with soil and water management. Farmers and ranchers, of course, depend on good soil and available water for their livelihood. Those who supply, service, and operate the machinery used to terrace, grade, or otherwise move the soil certainly have an interest in soil and water management. So do those who supply and service pipes and pumps that move the water. The landscaper and nursery worker spend much of their time devising means for managing the use of soil and water. The food-processing companies are very conscious of the need to conserve and control the use of water because they use a great deal of it.

Many times, of course, the farmer and

Figure 1-4. Irrigation equipment helps to avoid the crop damage caused by drought.

AN ORIENTATION TO AGRICULTURAL MECHANICS

rancher are unable to control what happens to the soil or whether water will always be available when needed. For example, not only were farmers and ranchers in parts of the Midwest plagued by the Great Depression of the 1930s; many were also forced to go out of business and leave their land because of the severe drought. Today's students will recall more recent droughts that have reduced crop yields and caused ranchers to sell cattle before they were ready for market. (See Figure 1-4.)

But, despite the fact that agricultural workers are often at the mercy of the forces of nature, more is being done today to manage and control soil and water than ever before.

Now heavy equipment is available for such purposes as terracing land, building waterways, and grading the land. There is also equipment to prepare land for cultivation that once would have been considered unsuitable for farm use. For example, swampy land can be drained, hills can be leveled, and huge boulders can be removed. Gullies and other depressions can be filled in easily to provide more cropland for production.

This ability to manage soil and water has become exceptionally important in recent years as the nation's population increased and the land area available for agricultural purposes decreased. Wise management of soil and water also has become more crucial as people discovered that the environment could be damaged through mismanagement.

Agricultural workers with responsibility for soil and water management need to be familiar with land-use planning. This involves making most effective use of the land and perhaps choosing among such alternatives as farming, highways, recreation, industry, and housing. Each year a land area the size of Delaware is taken out of production because it has been sold for housing or highway construction or has become too eroded for production. People working in soil and water management have to be able to deal with this kind of problem. Sometimes decisions have to be made on the basis of what is the most economical use of the land; but sometimes the decisions may be based on the need to preserve land for ecological balance or for its aesthetic value.

Workers in soil and water management have to know the procedures for leveling, running contour lines, and land measurement. They also need to be able to operate and maintain heavy equipment used by a soil and water conservation contractor.

Agricultural Construction and Maintenance

In almost all areas of agricultural industry there is a need for workers with the skills of carpenters, plumbers, masons, welders, metalworkers, and mechanics. Until 1963, these skills were commonly referred to as *farm shop skills*. But since these skills are employed in most aspects of agribusiness as well as on the farm, they are now referred to as *agricultural construction and maintenance skills*.

Workers in construction and maintenance need to know how to select and use a wide variety of tools and how to use them properly and safely. These tools may be used to build shelving, put up a new structure, install plumbing, place concrete footings and foundations, and do many other jobs in agricultural industry.

Food Processing

When nearly every family had a vegetable garden and raised its own meat, it was simple to get food from the producer to the consumers. The consumers were mostly the family members themselves. The family may have taken some products to a

Figure 1-5. Food-processing workers are sometimes asked to help evaluate equipment.

nearby town, where they would have been sold directly to consumers or perhaps to a grocer or butcher.

But all that has changed. More food is produced for more consumers, and they may live a continent or several continents away from where the products originate. Also, the products have to be packaged and transported in many different ways. Iceberg lettuce grown in California may be moved by refrigerated boxcar and sold in New York City; oranges grown in Florida may wind up as juice on tables in Seattle; wheat harvested in Kansas may travel by ship to India.

Thus the food-processing aspect of agricultural industry has become very big and very mechanized. There are machines for sorting and separating, washing, drying, cutting, heating, cooling, and packing. And often there are conveyer belts to move food from one operation to another. In fact, agricultural industry is so highly mechanized that food is often untouched by human hands from the time it is planted through the steps of processing.

Persons working in food processing should understand the principles and uses of various pieces of equipment, such as pumps, dryers, fans, and refrigerators. They also have to be able to install, operate, maintain, and repair the equipment.

Food-processing workers may be called upon to work with engineers and assist them in planning the construction and installation of equipment for transporting and processing materials.

Agricultural Mechanics: A Review

Agricultural mechanics is an important part of today's agricultural industry because the efficient production and processing of goods depend on mechanization. Therefore, the majority of workers employed both on the farm and in agribusiness need skills that enable them to operate, maintain, and repair machinery and equipment.

People who work in agricultural mechanics have their choice of many career

AN ORIENTATION TO AGRICULTURAL MECHANICS

fields. They may work outdoors or inside a shop or plant. Some jobs require physical strength, while others require that the worker mainly monitor automatic equipment and occasionally maintain and repair it. Some workers will use mostly hand tools, while others will use very sophisticated machinery powered by hydraulics or electricity.

People are no longer the machines in agricultural industry, doing many hard, time-consuming jobs. Today, with the proper skills, they are the managers of machines.

THINKING IT THROUGH

1. Using Department of Agriculture yearbooks and other historical references, trace the development of agricultural industry in the country and describe the impact of mechanization. The report you write may be used also as the basis for a speech to other students or in a local public speaking contest sponsored by the Future Farmers of America, 4-H, or a civic organization.
2. Using your own community, find out the types of jobs that are available in each of the six areas of agricultural mechanics.
3. Describe each of the six areas of training that are listed in agricultural mechanics. Give examples of what types of skills are required in each area.
4. Pick one of the six areas that interest you most. Visit people employed in that area and ask them about their jobs. You might ask them questions such as the following:
   (a) What are the future job opportunities?
   (b) What is the nature of the work?
   (c) What are the training requirements?
   (d) What are the physical requirements for the employee?
   (e) What are the career advancement possibilities?
   (f) What could someone expect to earn weekly at the start?
   (g) What is the salary scale or range?

CHAPTER 2

CAREER OPPORTUNITIES IN AGRICULTURAL MECHANICS

Employment reports show a steady decline in the number of persons employed in production agriculture. However, these figures don't tell the whole story.

There has been a decline, but it has been largely due to a reduction of unskilled farm workers who were replaced by machines that could do their jobs better, faster, and more economically. But there is room in the industry for persons with skills. A recent report by the U.S. Interdepartmental Committee on Employment Opportunities in Agriculture and Agribusiness states that only 29 percent of the employment needs in agricultural industry are being met by individuals who are trained to fill those needs. To put it another way, there are jobs for those who have the necessary training.

Increasing career opportunities in agricultural industry for the trained worker are predicted by the *Occupational Outlook Handbook* issued by the federal government.

As the technology of production becomes more sophisticated and widespread, the training requirements for farmers and ranchers increase. So do the training requirements for workers in agribusiness.

It is the time of the specialist, and the way to become a specialist is to learn the specialist's skills.

CHAPTER GOALS

In this chapter your goals are:

- To become familiar with the career opportunities available in the six areas of agricultural mechanics: agricultural power and machinery, agricultural structures, electrical power and processing, soil and water management, agricultural construction and maintenance, and food processing
- To identify specific jobs that require mechanical skills and abilities in the areas of production agriculture, agricultural supplies and services, agricultural mechanics, agricultural products, processing, and marketing, horticulture, renewable natural resources, and forestry

AN ORIENTATION TO AGRICULTURAL MECHANICS

- To identify some of the mechanical skills needed and the prospects for advancement in the various careers
- To list the personal qualities and educational requirements needed by workers in agricultural mechanics

Available Career Opportunities

Farm and Ranch Production

The unskilled farm and ranch worker is finding it increasingly difficult to get a job. Tomorrow's worker definitely will have to know more about tools and the operation, maintenance, and repair of machines and equipment.

When recently polled, a group of young farmers selected 73 tasks from a list of 103 tasks requiring mechanical skills as ones they regularly perform on their farm or ranch. These tasks range from servicing and adjusting the carburetor on a farm vehicle to installing electrical wiring and reconditioning tools.

The typical large farm and ranch uses a number of different vehicles. There are trucks of different sizes designed for particular uses. The most common is the pickup truck with a gasoline engine. It may be used for light hauling jobs, such as transporting parts, fuel, tools, lumber, and small amounts of feed and fertilizer. A 150-hp tractor with a diesel engine may be used for pulling a large chisel plow or other tillage tools. A van-type vehicle, a bus for transporting workers, and a flatbed truck also may be included in the fleet of vehicles kept by a farmer or rancher.

All these vehicles require people who know how to properly operate, maintain, and repair them. While a knowledge of power mechanics is essential, the worker also needs to be able to use welding and cutting equipment to help make repairs. And the worker may need to acquire some metal-working and carpentry skills in

Figure 2-1. There are many career opportunities in agricultural mechanics.

order to perform general maintenance jobs. For example, it may be desirable to construct shelves and storage bins in the shop, add grain sides to a flatbed truck, or build storage areas for a new combine or trailer.

Farms and ranches seem to add new machinery and equipment each year. One season the crop is harvested by hand or with simple tools; the next year a sophisticated piece of equipment is used to do the same job faster and more economically. In New Jersey, Rutgers—the State University—has developed new varieties of tomatoes that are suitable for machine harvesting. More and more farmers, therefore, are harvesting tomatoes by machine. Each year, farmers invest thousands of dollars in machinery that replaces hand labor.

The grain farms producing corn, wheat, and other grains for cash sale are highly mechanized to reduce labor requirements. Farms and ranches that produce milk, poultry, and livestock utilize mechanical equipment in most operations.

The *Occupational Outlook Handbook* describes mechanization in poultry and livestock farming. The raising of poultry in particular has become very mechanized, requiring only one or two workers to manage 50,000 young chickens. A suitable environment in the poultry house is maintained by using electric motors and controls. The feeding system is automatic, and the lights are controlled by a timer switch. The handbook states, "As in many farming enterprises, poultry farming requires specialized skills. The handling of the birds and of the mechanical feeding equipment requires specialized knowledge."

Referring to the cash grain farmer, the handbook advises future workers in that part of agricultural industry to have a very good knowledge of farm machinery operation and repair if they expect to be successful.

The worker skilled in agricultural mechanics knows how to operate, maintain, and repair the equipment and machinery used on the farm or ranch. The farmer or rancher who has invested $25,000 or more in a tractor will want a worker who can operate equipment properly and safely to prolong its useful life.

In the past, the typical farm and ranch consisted of a relatively few simple buildings in addition to the family homestead. Today's farms and ranches have many more structures. For one thing, now it is highly desirable to build structures to house all the machinery and equipment. Farmers and ranchers have found they save money by keeping expensive machinery in a shed. It reduces the weathering of belts, chains, rubber tires, and other parts. Also, the machinery that has been stored under cover generally is worth more money when traded or sold.

Structures also are needed to store the large quantities of fertilizer, pesticides, and other chemical products used by the modern farm and ranch.

The farm and ranch worker needs some carpentry and metal-working skills to help build and maintain structures.

Selling and Servicing Equipment and Chemicals

Farmers and ranchers annually purchase approximately 7 million tons [about 6½ metric tons (t)] of finished steel in the form of tractors, trucks, other vehicles, and building supplies. Farmers and ranchers invest almost $7 billion each year in new buildings, equipment, and machinery. Many more billions are spent for fertilizers, pesticides, and other materials essential to the operation of a farm or ranch.

When the huge investments in machinery, equipment, tools, vehicles, building materials, fertilizer, pesticides, and other materials by those engaged in ag-

Figure 2-2. Applying chemical fertilizer.

ribusiness are also figured, it is easy to see why there are many career opportunities for persons selling and servicing agricultural equipment and chemicals.

Persons with skills in agricultural mechanics are employed by suppliers to help install, demonstrate, and service equipment that has been sold and to repair equipment that is defective.

In some cases, the fertilizer supplier also provides personnel to run the equipment and machines for short periods of time. A farmer or rancher may contract to rent fertilizing equipment and an operator at the start of each planting season.

Often people are employed in agricultural supplies/services to operate and maintain feed blenders, mixers, grinders, crimpers, loaders, and forklifts.

As in most aspects of agricultural industry, the sales and service part also requires vehicles and storage facilities of different kinds. Therefore, there is a need for personnel qualified to operate and maintain vehicles and build storage facilities. The worker who is skilled in agricultural mechanics and also has a good business sense may very well wind up being an owner or part owner of a sales and service facility. And it is not unusual for workers with mechanical skills to start their own businesses where they contract their equipment and plow the land, plant the crops, fertilize, spray pesticides, and even harvest the products.

Agricultural Mechanics

The term *agricultural mechanics* is used here in a much more limited sense than when it applies to the overall body of knowledge discussed throughout this book. Here, agricultural mechanics refers to jobs requiring very specific mechanical skills. Persons who work on tractor engines and their helpers, "setup" persons who assemble irrigation systems and their helpers, and those who work on mechanical equipment in a feed mill or processing plant are examples of workers who are called agricultural mechanics.

Tractor dealerships and overhaul shops employ the largest number of agricultural mechanics. There are people who sell parts, overhaul tractors and equipment, and work on machinery.

Most persons starting off in such shops are employed first as helpers. Then they progress according to their abilities and additional training. For example, most tractor mechanics attend special factory training programs to learn how to maintain and operate the equipment.

Food and Nonfood Processing

As the nation entered its third century, there were more than 2 million persons employed in some aspect of food and nonfood processing and handling. Food products processed include meat, vegetables, poultry, eggs, dairy products, fruits, and some cereal grains. Nonfood products processed include wool, cotton, tobacco, lumber, and such animal feeds as corn, soybeans, and cottonseed.

It may come as a surprise, but approximately two-thirds of the food products in today's supermarkets were not available just 15 to 20 years ago. Examples of such products include most of the freeze-dried foods, the so-called instant products (mashed potatoes, soup, and coffee), frozen dinners, and many other dried or frozen foods. Today 75 percent of food products are processed in some way. That wasn't true when your parents were your age or younger.

To reduce the costs of production and to increase the volume of production, agricultural industry has mechanized most of the steps in the processing and handling of both food and nonfood products.

The processing and handling equipment may be as simple as a conveyer belt or as complex as huge cookers that prepare foods for canning and freezing.

Many thousands of persons are employed to install, operate, and maintain the dozens of different pieces of equipment and machinery. These are well-paying jobs, because the people must be highly skilled in agricultural mechanics. They have to be familiar with the equipment, of course; but they also may be called upon to use such skills as welding, carpentry, and metal working to keep the processing plant running.

A worker may have to use a welding torch to repair equipment that has broken down, and carpentry skills may be required to construct platforms and equipment supports (see Figure 2-3). Since most of the equipment is made of metal, the worker who has metal-working skills will be able to repair and replace parts more easily.

Since food and nonfood processing also involves storage and transportation, persons are needed who can install and maintain storage facilities as well as people who can operate and maintain a variety of vehicles. These include forklifts and small tractors to move a pallet loaded with boxes or crates. A *pallet* is a platform that can be used to move goods from one place to another.

Workers may start as assistants to persons with more skill and experience,

Figure 2-3. Mechanics skills are needed to keep agricultural equipment in good working order.

but they can advance to positions with better salaries. In large processing plants, for example, it is not uncommon to have vice presidents or assistant vice presidents in charge of plant operations.

Horticulture

Horticulture includes the production and merchandising of fruits, vegetables, flowers, nursery stock (shrubs and trees), and turf grass. Also included is landscaping, the installation and care of golf courses, and greenhouse operation.

Many persons in agricultural industry consider the production of vegetables to be farming. So the section on farm and ranch production also could apply to the raising of vegetables.

AN ORIENTATION TO AGRICULTURAL MECHANICS

Persons engaged in horticulture sometimes may be known as specialty farmers or growers. Their specialty may be oranges, pecans, gladiolas, holly trees, grapes, or other fruits, vegetables, or ornamental plants.

The horticultural aspect of the industry is also very mechanized. Crops are being picked by machine that only a few years ago were picked by hand.

The person with mechanical skills may get a job spraying trees or fields with herbicides or pesticides. Special training may be required, however. Many persons using herbicides and pesticides must be licensed to do such work. The self-employed person may own and operate various kinds of spraying equipment that are available to local orchards, vineyards, and specialty farms.

When workers in nurseries were questioned recently, they reported that 11 of 53 skills necessary in their jobs were in agricultural mechanics. These skills ranged from care and use of hand tools to the daily maintenance of tractors and the installation and maintenance of irrigation systems.

The person employed in *floriculture*—the commercial production of flowers—must be skilled in basic carpentry and maintenance. The carpentry skills are needed, for example, to keep greenhouses in good repair and to build platforms and flats for plantings. Greenhouses are built to very strict specifications to control temperature, light, and humidity. Extensive control systems must be installed and maintained to ensure automatic watering and ventilation.

The turf grass business has expanded in recent years. In this business, special, high-quality grass is grown and sold to homeowners, golf courses, industries, and others. The turf grass is laid and maintained usually by the same people who grow it. Workers have to know how to operate such equipment as spriggers (to plant grass shoots) and sod cutters.

The landscaping business also is highly mechanized, and workers have to use and maintain such equipment as lawn sweepers, power rakes, fertilizer spreaders, sprayers, and aerators (to poke holes in the soil for air and moisture).

Many persons who have mechanical skills as well as other knowledge about the field eventually become independent nursery owners, turf grass producers, and landscapers. While there are risks in these

Figure 2-4. Putting down turf grass.

fields as in any business, there also are profits to be made.

Arborculture is a part of ornamental horticulture, and it has to do with the care of trees. Workers in this business must be able to operate and maintain such equipment and tools as sprayers, power saws, pruners, branch shredders, and power winches and hoists for lifting and moving whole trees and large sections of trees.

Renewable Natural Resources

People responsible for the management and preservation of such resources as land and water need many mechanical skills.

If, for example, a person were employed either by a private company or by a government agency to maintain renewable natural resources, that person would have to perform basic equipment maintenance and operate earth-moving machinery used in digging trenches and building ponds and terraces.

Also there are jobs in this field for conservationists, engineering technicians, conservation contractors, fish and wildlife officers, and park employees. While these positions require different skills and often advanced education, training in basic agricultural mechanics skills is desirable.

In addition to using vehicles, earth-moving equipment, and tractors, persons in these jobs often operate power tools and do incidental carpentry.

Forestry

People are employed in the field of forestry to plant, manage, and harvest trees. There are also jobs in preserving forests both as wilderness and for recreational purposes.

Employment opportunities exist in private lumber companies; in federal government agencies such as the U.S. Forest Service, National Park Service, and Bu-

reau of Land Management; and with local and state agencies.

Like the other fields mentioned, forestry is highly mechanized and requires personnel who have mechanical skills. These days very few jobs are performed that exclusively use physical strength. Giant machines are now available that can shear off a tree at ground level and remove branches in one operation. People employed to use these machines must know how to maintain and repair them.

Because of the mechanization in agricultural industry, today's employees need to know more about hydraulics, engine maintenance, welding, and equipment repair than about swinging an ax and otherwise using muscle power.

Tractors are used to pull planters that set out seedlings, and power sprayers are used to protect the trees from insects and disease. Machines called *skidders* load logs on large tractor trailers for shipment to saw mills (see Figure 2-5).

Figure 2-5. A large truck taking logs to a sawmill.

AN ORIENTATION TO AGRICULTURAL MECHANICS

Career Opportunities: A Review

Agricultural industry may be larger than you thought at first. Usually, when agriculture is mentioned, many persons think only of a farmer with straw hat and pitchfork.

Today's farmers and ranchers must be good managers who know how to use mechanization to their advantage. They make maximum use of workers with specialized skills. But workers with mechanical skills are also needed off the farm in agribusiness, food-processing plants, greenhouses, nurseries, landscaping companies, soil conservation operations, and the lumbering business and forest service.

The skills that are needed to operate, maintain, and repair modern tools and equipment are required in all aspects of agricultural industry.

THINKING IT THROUGH

1. List two job titles that require mechanical skills in each of the following occupation areas:
   (a) Production agriculture
   (b) Agricultural supplies/services
   (c) Agricultural mechanics
   (d) Agricultural products, processing, and marketing
   (e) Horticulture
   (f) Renewable natural resources
   (g) Forestry
2. Identify the types of skills that are necessary for three of the jobs that interest you most. Identify skills according to the categories of agricultural power and machinery, agricultural structures, electrical power and machinery, soil and water management, or agricultural construction and maintenance.

CHAPTER 3

PERSONAL CHARACTERISTICS FOR SUCCESS IN AGRICULTURAL MECHANICS

''Know thyself'' said the ancient philosopher Socrates, and his advice is still good. Persons planning their future career must take stock of themselves: their personal strengths and limitations, the extent of their knowledge and skills, their aptitudes, their natural mechanical and intellectual abilities, the state of their health, and how they relate to other human beings and their surroundings.

It's not always easy to take stock of yourself. To begin with, it's hard to be objective. But a candid examination of what and who you are is essential to success in life.

In this chapter there are ways for persons to measure themselves as they consider a possible career in agricultural industry. Also discussed are the basic knowledge, natural ability, skills, and physical and mental qualities needed to do well in agricultural industry.

CHAPTER GOALS

In this chapter your goals are:

- To describe the five steps involved in making a career decision in agricultural industry
- To chart a career direction for yourself
- To assess your own qualifications for the future you want
- To list physical and mental abilities and mechanical skills needed to enter and advance in agricultural industry

How to Make Career Decisions

Some persons decide what they want to do with their life while they are still in junior high school. Others make career choices in high school as a result of courses they take, teachers they have, and other influences. Still other persons decide what they want to do after they get out of high school and work a few years.

There is nothing wrong with changing your mind about a job or career field. However, it can be helpful to follow some simple steps in making a career choice.

Observe the World of Work

A key to career planning is to observe people at work. Not only is it a way to see how jobs are performed and what skills are required to do them, but also it is a way to find out what jobs seem interesting and satisfying and what jobs don't.

Consider the different ways two students, Jim and Gene, observe the world of work while on a visit with their parents to the local machinery dealer.

Jim watches the five persons working in the store. He notices that each person is doing something different. One worker is taking inventory of parts. Another is examining how a spreader works. Two other workers are going outside with a customer to point out what repairs were made on a tractor. The fifth employer at the moment is waiting on Jim's parents.

Shortly after Jim leaves with his parents, Gene walks in with his father. While his father talks to one of the workers, Gene heads for the soda machine. He buys a can of soda and looks over the posters on the company's bulletin board.

If both Jim and Gene later apply for a job at the dealership, Jim probably will be the one to have a better idea of what he can expect from the job and how well suited to it he is. Gene is more likely to have unrealistic expectations about the job. And unrealistic expectations can lead to disappointment and frustration.

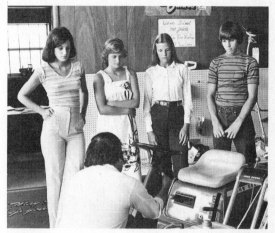

Figure 3-1.   A key to career planning is to observe people at work.

Analyzing the Observations

Jim not only observed the people at work in the dealership, but also analyzed, or thought about, what each person was doing. He determined that the one worker was a machinery parts clerk, another probably was a setup mechanic (assembling the spreader), the two workers going out to demonstrate the tractor were a mechanic and helper, and the worker waiting on his parents was a salesperson.

He also observed the workers' ages. He noticed that the machinery parts clerk and the mechanic's helper were the youngest workers. He decided they were probably the least experienced and that a machinery parts clerk and mechanic's helper probably are entry-level jobs in this dealership (a part of agribusiness).

Making a Decision

Even after observing and analyzing, some people put off actually making a decision. Some may have a fear of making the wrong choice. Others are procrastinators—they always put off things until tomorrow.

After analyzing several related jobs—called *job clusters*—decide what appeals to you and what is a realistic goal to set for yourself. Then begin the steps toward career preparation. Most careers are a series of jobs with the success in one job leading to another job, usually with more responsibility and a better salary.

Taking Action

After the observations are made and analyzed, it's time to act. Jim has figured out that the mechanic's helper is the place to start if one wants to become a tractor mechanic. He likes engines and believes he has some ability to work on vehicles such as a tractor. He decides, therefore, that he might want to become a tractor mechanic. He talks about his decision with his vocational agriculture teachers at school.

Assuming Responsibility

Even though you get assistance from your vocational agriculture teacher, school counselor, parents, and friends, you are the person responsible for your career. After observing the world of work, analyzing observations, making a decision, and taking action, you are ultimately responsible for the outcome. By assuming this responsibility, you become the manager of your career.

Most decisions can be changed if the situation is altered. Keep your options open as much as possible. And keep in mind where the responsibility lies.

Planning for the Future

Jim plans his high school curriculum carefully, even though later he may change his mind about becoming a tractor mechanic.

The teachers suggest courses that will be good preparation for a job as a tractor mechanic, but they also suggest courses that will give Jim a sampling of other agricultural mechanics skills—just in case he does change his mind. For example, they recommend some business courses that could prepare Jim for sales work in agribusiness and for running a dealership.

Taking Advantage of Opportunities

Jim will be smart if he takes advantage of supervised occupational experience programs sponsored by his school in cooperation with local agricultural industry. By working under supervision, he can learn

Figure 3-2. The FFA offers many learning situations in agricultural mechanics.

AN ORIENTATION TO AGRICULTURAL MECHANICS

more skills needed to enter his chosen career.

Also, Jim can see what working conditions are really like. The more he learns about the requirements of being a mechanic's helper (or other job), the better prepared he will be to meet those requirements.

Other opportunities for learning are the activities sponsored by the Future Farmers of America (FFA). FFA contests in agricultural mechanics, for example, enable young persons to perfect and test their abilities and skills. Many people have succeeded in agricultural industry because of their initial experience with FFA.

A Good Education Is Necessary

Agricultural industry, as well as every other field, requires workers to be competent in basic skills. That means that a person must be able to read well enough to learn from what others write and to write well enough to communicate adequately with others. Simple arithmetic and an understanding of metrics are also a basic requirement.

Today's tractor mechanic must be able to read technical manuals, make very precise measurements, and write instructions for others to follow.

Beyond this basic foundation, a person's educational background can be as varied as the demands of the career and personal desires and interests. If one wants to advance in agricultural industry, most experts recommend a background in the principles of production (plant and animal science), leadership, management, marketing, science, and technology.

The investments in production agriculture and agribusiness are now so great and the applications of science and technology are so complex that it can be very costly to make a mistake. Therefore, young people who want to progress to higher levels in agricultural industry should consider a post-high school education. Some of the courses that might be helpful are chemistry, botany, economics, and soil and water conservation.

Mechanical Ability Is Important

Because agricultural industry is so mechanized, nearly every worker has to have *some* mechanical ability and skills, and many workers need considerable mechanical skills for their job.

Earlier in this chapter, self-evaluation of strengths and limitations was discussed. One way to evaluate strengths and limitations is through aptitude tests that provide information on how much ability a person already possesses for doing certain kinds of tasks. The test results should not be considered absolute, but they are a guide for the student and teachers.

People who are limited in some skills needed for a particular career in which they are interested may be able to acquire or improve those skills in school. For example, a student with limited public speaking skills may participate in creed speaking or public speaking contests in the FFA. A student with limited mathematics skills can improve them through record-keeping (personal records and financial records). Supervised occupational experience (SOE) programs give students many opportunities to improve skills and to gain valuable work experience.

But if the tests show that a person has an extremely low potential for certain skills, the person might want to consider going into some other area of agricultural industry. Teachers and counselors can help in this evaluation.

Regardless of the career in agricultural industry, most persons will need training in the safe operation and maintenance of vehicles and equipment. Most areas also require ability in basic mechanical skills.

Good Health Is Also Important

Good health is something no person can take for granted. Employers want their employees to be in good health (except for occasional short illnesses and emergency situations). They want employees to make a maximum contribution to the business. Employees are, or should be, concerned about their health for the same reasons. Also, employees with poor attendance records caused by minor or major health problems are often those persons fired when cutbacks are necessary, or they are overlooked for advancement in the field.

Some jobs in agricultural industry,

such as feed mill and equipment operators, require more physical stamina, dexterity, and muscle coordination than those in many other fields. While a person who has suffered from rheumatic fever might do well in an indoor job requiring little exertion, that person would not fare well in an outdoor job requiring the lifting, pulling, and pushing of heavy objects. Persons with severe allergies to certain kinds of pollen and insects probably should not consider outdoor jobs that will bring them in close contact with pollen and insects. Of course, many allergies can be successfully controlled by taking medication as prescribed by a physician.

Persons with weakness in their arms, legs, and back also should consider how well they can perform certain jobs. Many problems can be corrected, of course. A person who lacks sufficient strength in the arms and legs may be able to do certain exercises that can help. These should be

Figure 3-3. Some jobs require good physical stamina.

AN ORIENTATION TO AGRICULTURAL MECHANICS

done only after consultation with a doctor, therapist, or physical education specialist.

The fact that persons have physical impairment does not automatically disqualify them from most jobs. It is not uncommon these days for persons with artificial limbs to perform very strenuous activities with nearly the same ease as people with natural limbs.

Consider Working Conditions

There are some jobs in agricultural industry that keep employees inside much of the time, such as working in a processing plant, welding shop, or agricultural supplies store. Persons contemplating certain jobs should be honest with themselves. They should not take outdoor jobs when their preference is for inside work, and vice versa.

Many persons who work in agricultural industry have to be prepared to work during bad weather and under conditions that aren't always pleasant. A field mechanic for a tractor dealer, for example, may have to lie in the mud to make repairs on a tractor. An equipment operator might have to drive a bulldozer in the hot sun. The logger may have to work in the forest when the temperature is far below freezing. A custom farmer may have to operate haying equipment in a swampy area infested with mosquitoes.

These descriptions are not meant to discourage anyone from taking a job outdoors. Just the opposite is true. The purpose is to help people prepare for a variety of agricultural jobs and to be realistic about each job's requirements.

Getting Along With People

Sometimes employees in agricultural industry work alone, but most of the time they are working with other persons. In many cases, the employee has to rely on the efforts of coworkers. Many harvesting operations, for example, depend on very close teamwork between persons operating different machines. The worker using carpentry skills to build a structure must look to other workers to assist in the effort. People working in sales and service must cooperate closely with one another to guarantee the best service to customers.

Be Honest with Coworkers

Agricultural industry places great stock in honesty and integrity. People must be able to trust the worker who says a piece of equipment has been repaired and is now safe to use. The public must be able to rely on the quality of food raised and processed by the industry. If workers say they can assemble the parts necessary for an irriga-

Figure 3-4. The ability to work with others is important in almost any job.

tion system to bring water to dry crops, many people will put their faith in that promise.

Be Even-Tempered

People should be emotionally stable when dealing with coworkers and other persons. People who lose their temper frequently or become depressed and "moody" may be a problem to others and a liability to themselves.

Young persons who are active in the FFA generally learn how to work and get along with people of different temperaments, interests, and backgrounds. Such experiences can prove invaluable later when it comes time to work with people on the job. Active involvement in the FFA helps develop leadership and human relations skills.

Desirable Personal Characteristics: A Review

It is very important for persons planning careers in agricultural industry to understand first their own strengths and limitations and then how they can capitalize on these strengths and overcome the limitations.

Making decisions and successful career choices depends on purposeful observation of the world of work and careful analysis of those observations. And each person must be prepared to accept the consequences of decisions and actions taken.

As students plan for their future, they must take stock of themselves and take advantage of educational opportunities and work experiences. The investment made now in time, energy, and thought can pay very real dividends in the future.

THINKING IT THROUGH

1. List and describe the five steps involved in making a career decision in agricultural industry.
2. Why is a good education necessary for employment in tomorrow's agricultural industry? What constitutes a good education?
3. List three favorite courses that you have studied in school. Why did you like them?
4. Of the courses you have studied, which did you like least? Why?
5. Thinking about the courses you preferred, identify two jobs in agricultural industry which might interest you. How would the courses you preferred help prepare you for those jobs?
6. Describe the mechanical skills needed to enter and advance in the two jobs you listed.
7. Describe the personal skills needed to enter and advance in the two jobs you listed.

UNIT II

SKETCHING, DRAWING, AND PLAN READING

COMPETENCIES

Competencies	PRODUCTION AGRICULTURE — Livestock farmer or rancher	AGRICULTURAL SUPPLIES/SERVICES — Supplies salesperson	AGRICULTURAL MECHANICS — Equipment setup person	AGRICULTURAL PRODUCTS, PROCESSING, AND MARKETING — Fruit and vegetable processing worker	HORTICULTURE — Landscape supervisor	FORESTRY — Forester aide	RENEWABLE NATURAL RESOURCES — Soil conservation technician
Identify, select, and use sketching and drawing equipment	Very Important	Important	Important	Important	Very Important	Very Important	Very Important
Sketch a plan	Very Important	Very Important	Very Important	Important	Very Important	Very Important	Very Important
Read and interpret a schematic drawing	Very Important	Very Important	Very Important	Important	Important	Important	Very Important
Read an orthographic drawing	Very Important	Very Important	Very Important	Important	Very Important	Very Important	Very Important
Hand-letter a sketch or drawing	Important	Very Important	Important	Important	Very Important	Very Important	Very Important
Figure and prepare bills of materials	Very Important	Very Important	Very Important	Very Important	Very Important	Very Important	Very Important

 Very Important   Important   Not Important

There seems to be a sign for almost every occasion. Some signs get their message across very simply: "Speeders Lose Licenses." Other signs, such as the bumper sticker that says "Give an Animal a Brake," may state their message more subtly and cleverly.

For many years a favorite sign has been the one reading "Plan Ahead." The message may be subtle, but it's certainly to the point. One can't just talk about planning and hope everything will work out. To plan means to think and prepare carefully and thoroughly before undertaking an action.

People make many plans, some simple and some complex. Students may make a plan for a class party—where it will be, who will bring what, the time, etc. A high school program of study also should be planned—what courses to take and when to take them. This planning can be critical. For example, if students have taken the wrong courses in high school, they may not be able to enter the agricultural field of their choice. And some students take courses in high school that do not prepare them for later study at a technical school or college.

Planning is crucial in all agricultural career fields: production agriculture; agricultural supplies/services; agricultural mechanics; agricultural products, processing, and marketing; horticulture; renewable natural resources; and forestry. Being able to plan correctly is just as important as having mechanical skills and having the desired personal characteristics for the job.

One can't construct a project such as a livestock trailer, for example, without first thinking about the kind of trailer desired, deciding what its purpose will be, and figuring how much material and how many hours will be needed to complete the project. Then the concepts must be turned into a detailed plan called a *blueprint* that a welder can follow exactly. The blueprint ensures that the completed trailer will be built to the correct size for the loads it must carry.

Simpler projects may need only a sketch for the concept to become a reality. One might sketch where benches will go in a greenhouse and the dimensions of the aisles between the benches. A sketch might be drawn to show where the electricity will enter the greenhouse and how the electrical circuits will be routed inside the structure.

The skills to be learned in this unit are drawing sketches, making orthographic drawings, and reading plans prepared by others.

Upon completion of the unit, you should be able to understand basic graphic presentations and the procedures and symbols used in shop blueprints and sketches. Also you should be able to do orthographic drawing, using a pencil to make whatever lines and shapes are necessary to show the dimensions of an object.

You should know how to do simple lettering that will give messages and instructions to workers using blueprints and orthographic drawings. And you should be skilled in filling out a purchase order for materials and supplies.

CHAPTER 4

THE ELEMENTS OF A PLAN

Many members of the Future Farmers of America (FFA) begin their supervised occupational experience with a livestock project. As the members become more knowledgeable and skilled, their responsibilities increase. They may be required, for example, to build a set of working lots or corrals where beef, swine, or sheep can be isolated and treated when they get sick.

Before these can be built, however, it is necessary to draw a simple sketch that shows where everything should be located, including a squeeze chute and a loading chute for the livestock to be routed into and through the lots or corrals. The sketch helps the FFA members build corrals that are both efficient and economical.

Several companies specialize in the construction of agricultural buildings. Workers employed by these companies do everything required by the project, from preparing the building site to installing the equipment after the building is completed. This start-to-finish operation is called a *turn-key job*. Many livestock structures such as turkey and chicken houses, swine-feeding barns, and milking parlors are turn-key jobs. That is, an outside company not only plans and builds the structure but also puts in the equipment and gets it ready for operation.

The people involved in such turn-key operations usually work from a set of plans. These plans generally are called *blueprints*, because of the blue tint of the copies reproduced from the original plans. They contain all the details necessary for each phase of the job. Blueprints usually are prepared by an engineer, architect, or draftsperson.

Both the simple sketch that an FFA member draws and the complicated blueprints used by a turn-key company are visual plans. However, most of the visual plans needed in agricultural industry are sketches. In those few cases where complex blueprints are required, persons with special training usually are required.

CHAPTER GOALS

In this chapter your goals are:

- To identify the three common types of plans and describe how each might be used
- To identify the elements of a sketch and the equipment necessary to draw one
- To letter instructions and dimensions that can be used in labeling sketches and for other purposes
- To identify and describe the correct use of drafting symbols

Sketches

A *sketch* is a visual plan. It is intended to turn ideas into finished projects. Sketches are used by many agricultural workers. Persons who need to draw and read sketches include the landscape worker, food-processing plant mechanic, farm equipment setup mechanic, soil conservation aide, bulldozer operator, and livestock-feeding equipment salesperson.

The sketch communicates an idea or mental picture of a project to another person, so that the person can see the project as clearly as the one who first envisioned it. The sketch usually is a simple freehand drawing, but it is not amateurish or careless. It is not the polished drawing an architect or draftsperson might complete, but it is done neatly and carefully. A sloppy drawing may produce a finished project that does not resemble what the planner had in mind.

The sketch may be drawn primarily to show what something is supposed to look like when it is completed. The sketch also may show details and dimensions. For example, a farm building construction worker may want to build a small, portable shed. The worker would start by drawing a simple sketch to show how the shed will look when it is finished. The worker also may develop more detailed sketches to show the exact dimensions of the shed.

Many Workers Draw Sketches

The examples of workers in various agricultural career fields who need to know how to sketch plans are numerous.

A farm or ranch worker may be asked to sketch ideas for a new building to handle livestock, and the employee of an agricultural feed store may be asked to sketch the location of electrical connections in a switch used in a feed-mixing mill. The mechanic who has to assemble equipment or machinery may sketch a plan to help put the right piece in the right place. A sketch is needed by a nursery worker who must develop a landscaping plan. And a person employed in land management may draw a sketch of where to locate drainage or irrigation ditches. Regardless of the career field in agricultural industry, there is a need for sketching.

Simple Tools Needed for Sketching

Sketching is not complicated. It does not require much equipment or technical know-how on the part of the person making the sketch. The worker needs just a pencil, eraser, and paper for most basic sketches. In some cases, the worker may want to use a straightedge or template (pattern).

The pencil used should have hard lead. The best pencils are those marked HB, F, or H. It is very important to keep pencils sharpened. Otherwise, the lines become too wide, or they may smudge. A firm rubber eraser is needed to make changes in the sketch.

The paper that works best for sketching is *grid paper* (often called *graph paper*). The paper consists of a series of lines that cross one another to form a grid. Grid paper can be obtained in various sizes. Some paper contains four squares per inch [2.54 cm] while others have five, eight, or some other number of squares per inch. A standard grid paper for many projects is one that contains eight squares per inch. Such paper is desirable when dimensions are figured according to the existing U.S. Customary System of measurement that uses fourths, eighths, sixteenths, and thirty-seconds of an inch.

The grid paper makes it much easier to draw objects in their proper proportion. For example, the worker may want to show the floor layout of a farm shop. The floor is 28 × 40 ft [8.5 × 12.2 m]. The worker decides to make ¼ in [0.6 cm] represent one ft [0.3 m]. On the grid paper, therefore, the outline of the shop is reduced to 7 × 10 in [17.8 × 25.4 cm]. This is called *drawing to scale*. It is not even necessary to use a ruler. If eight squares equal 1 in, then one dimension will consist of 56 small grid squares and the other dimension will consist of 80 squares. The sketch now shows the floor of the shop with its proper proportions.

Often grid paper is colored light green, orange, or blue, although it also can be purchased in white. Grid paper generally is available in the supply section of any office supply store.

If grid paper cannot be purchased, any good-quality paper will do. Some very acceptable sketches have been drawn on an inexpensive pad. However, onionskin paper or any other very thin paper that will make erasures difficult should not be used. Also, it is hard to draw on some coated papers.

Composing the Sketch

Sketches are composed of a series of straight and curved lines. Without a straightedge, it is sometimes hard to draw a long, straight line. But it's quite easy to draw short, straight lines (see Figure 4-1).

A compass or template may be used to draw a circle, but a circle can be drawn on grid paper by using a series of very short lines. With practice, anyone can learn how to avoid bulges or flat spots on the circle.

The sketch should show the basic characteristics of what is being presented. For example, the soil conservation technician may need to sketch a field that is irregular in shape. The irregularities need to be shown, even if the drawing is not exact. If a stream crosses the property, then the stream should be positioned on the sketch

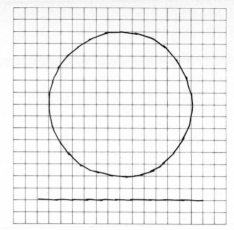

Figure 4-1.   Drawing a straight line and circle on grid paper.

in relation to where it is located on the field. Other landmarks also should be drawn proportionally. It could prove both embarrassing and expensive if the sketch of a field where a pond is to be built did not take into account a gully, outcropping of rocks, or other obstructions.

Schematic Drawings

Schematic drawings are diagrams. Their primary purpose is to show the relationship of one object to another. Symbols are used to represent the objects that are included in the actual project. An important difference between the schematic drawing and the sketch is that proportions are not particularly important in the schematic drawing.

An agricultural worker installing an electrical system generally will use a schematic drawing to show the relationship of the components of the circuit (see Figure 4-2). The person working in landscaping may use a schematic drawing to show the positioning of different trees and shrubs. The employee of an equipment service agency may use a schematic draw-

SKETCHING, DRAWING, AND PLAN READING

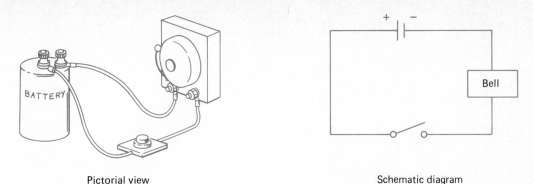

Pictorial view   Schematic diagram

Figure 4-2. Schematic drawing showing the components of an electrical circuit.

ing to show the relationship of grain-handling facilities and equipment. Figure 4-3 shows how a farm machinery worker used a schematic drawing to show the relationship of hydraulic equipment used on an agricultural machine.

Orthographic Drawings

The orthographic drawing is more advanced and complex than a sketch. Therefore, it requires more skill to make a good orthographic drawing.

This kind of drawing gets its name from the prefix *ortho*, meaning at right angles. In an orthographic drawing, the different views of the object are all drawn at a right angle (90°) to the surface of the object (see Figure 4-4).

Orthographic drawings show all sides or views of an object, and they show them to scale. That means that the dimensions of the drawing are in exact proportion to the actual dimensions of the object. A 16 ft [4.9 m] trailer drawn to a scale where ½ in [1.3 cm] equals 1 ft [0.3 m], would appear 8 in [20.3 cm] long in the drawing.

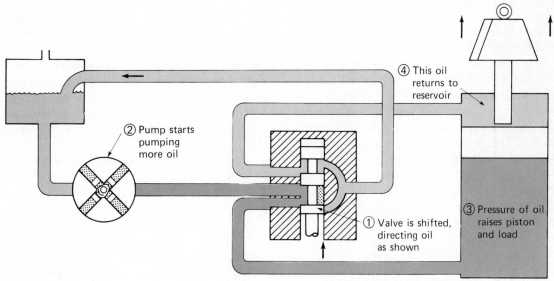

Figure 4-3. Schematic drawing of a closed-center hydraulic system.

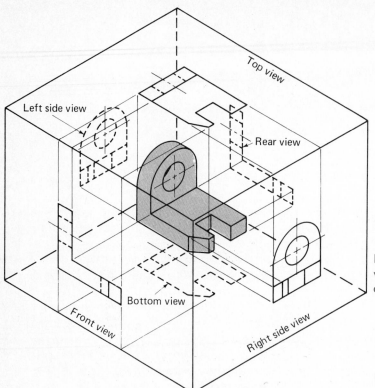

Figure 4-4. The principal views of an orthographic drawing.

Every dimension can be seen and measured. It is almost as if the object were enclosed in a glass box, allowing the viewer to see everything clearly (refer to Figure 4-4).

It is possible in an orthographic drawing to show six views of an object: *front*, *rear*, *right side*, *left side*, *top*, and *bottom*. In many cases, however, it is only necessary to show three views: front, right side, and top. The right side of an object is always the right side as one looks at the object from the front. The other views—left side, rear, and bottom—may not be shown unless they differ importantly from the three fundamental views.

An orthographic drawing uses symbols, sometimes called *conventions*, to describe the object. For example, a solid line is used to show any dimension that normally would be seen by someone looking at the front of the object. A broken line made of short dashes is used to show dimensions that normally would be hidden from view. If a center line is used in the drawing, it consists of short and long dashes. Generally a center line is used only when one wishes to show the center of a round shape.

While the orthographic drawing represents the views as arranged in Figure 4-4, when the views are actually drawn on paper, they are arranged differently. It is almost as if one were to "unfold" the arrangement in Figure 4-4, as shown in Figure 4-5. Once the views have been "unfolded" and laid flat, they are shown as they are seen in Figure 4-6.

SKETCHING, DRAWING, AND PLAN READING

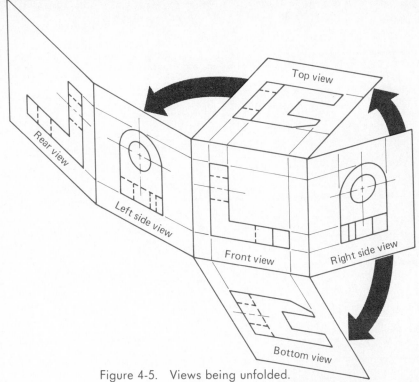

Figure 4-5. Views being unfolded.

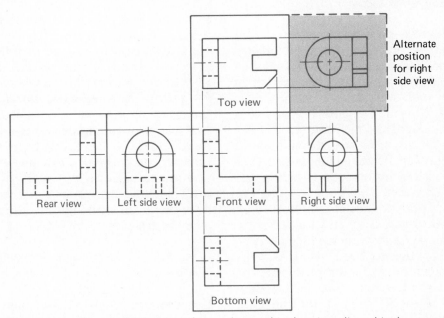

Alternate
position
for right
side view

Figure 4-6. The principal views of an orthographic drawing aligned in the standard arrangement.

THE ELEMENTS OF A PLAN

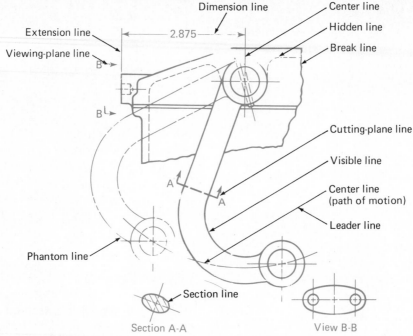

Figure 4-7. Drawing symbols used to show the dimensions of an object.

How to Show Dimensions

When showing dimensions on an ortho-graphic drawing, the worker should use dimension lines (see Figure 4-7). Some-times there isn't room on the drawing to show a complete object. In such cases, a break line is used. This line indicates that part of the object has been omitted. *Leader lines* are used to refer to a particular point on the drawing. Other symbols (conven-tions) are used by engineers, architects, and draftspersons, but they will be ex-plained in more advanced courses.

Dimensions are given in exact mea-surements, and they should all be given in one system (the metric or the U.S. Cus-tomary System). A single drawing, for example, should not use both feet and me-ters. However, if one is using the U.S. Customary System, both feet and inches may be used to be exact in the mea-surements. It is unlikely that all dimen-sions would measure exactly in feet, and using only inches would be very cumber-some in drawing large objects. If the metric system is used, meters and centimeters may be combined.

Although some agricultural workers are called upon to do an occasional ortho-graphic drawing, more workers are re-quired to read and work from them. A number of orthographic drawings are commissioned from architects and draftspersons. However, instructions for developing an orthographic drawing are given in Chapter 6.

Lettering

Good lettering makes labels and instruc-tions on sketches and orthographic draw-ings easier to read. Lettering that is neatly

SKETCHING, DRAWING, AND PLAN READING

Vertical letters

ABCDEFGHIJKLMNOP
QRSTUVWXYZ&
1234567890

abcdefghijklmnopqrstuvwxyz

—Height of general
drawing lettering

$\frac{2}{3}$ Height of general
drawing lettering

Inclined letters

ABCDEFGHIJKLMNOP
QRSTUVWXYZ&
1234567890

abcdefghijklmnopqrstuvwxyz

—Height of general
drawing lettering

$\frac{2}{3}$ Height of general
drawing lettering

Figure 4-8. Single-stroke commercial lettering style.

done also gives the sketch a more professional, finished look. The person who becomes accomplished in lettering can also use the skill to make signs. It could be a simple "Keep Off" sign for newly seeded grass, or it could be a sign near equipment and machinery that gives important information about its operation.

Using Different Lettering Techniques

There are many styles of lettering and many lettering techniques. When one has mastered a basic style or technique, it is possible to experiment with others.

The most common style of lettering is *single-stroke commercial* (see Figure 4-8). This style is easy to read and is fairly easy to draw. The style consists of both uppercase and lowercase letters (capital and

small letters). There are variations of the style. One consists of slanted letters, called *italic*. Since drawing italic letters is more difficult, the regular letters, called *roman*, should be mastered first.

To make sure letters are properly aligned, guidelines should be used at the top and bottom of the letters. The guidelines will help keep the letters the same size and in a straight line.

Any good paper can be used for lettering, and a straightedge can be employed to draw the guidelines. Grid paper also can be used for lettering (see Figure 4-9).

Practice Lettering

The best way to practice lettering is to draw letters. The capital letter shown in Figure 4-8 should be studied before the practice begins. The stroke pattern for each letter should be examined. The next step is to practice each stroke. Individual letters should be drawn before words or sentences are attempted. The order of strokes and the proportions of the letters should be learned through practice until you can draw them quickly and accurately. Numerals and fractions also need to be practiced.

Commercial draftspersons employed in agribusiness sometimes use lettering guides or scriber templates for more professional quality in lettering. Several kinds

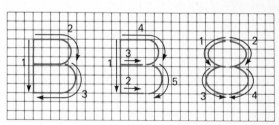

Figure 4-9. Sequence of strokes used to form letters and numbers.

Figure 4-10. Using a Leroy lettering guide.

of lettering guides are available. The most common scribers are the Leroy lettering guide (see Figure 4-10) and the Wrico Universal lettering guide.

Also available in many stationery and art supply stores are sheets of transfer or rub-on letters. This is a simple but expensive process to use. Both capital and small letters are transferred from the sheet to the paper by rubbing the letters with a pencil or stylus. Transfer lettering comes in a great variety of type styles.

For most of the lettering they need to do, however, the majority of agricultural workers find they do just as well drawing letters by hand.

Elements of a Plan: A Review

A sketch is a simple visual plan that helps workers in all agricultural career fields develop and illustrate ideas and relationships. The sketch helps translate an abstract idea or concept into a concrete view of the shape, size, and layout of an object. Although the sketch is simple, it is not done carelessly. The sketch should illustrate the proportions of the object and, in many cases, give the details and dimensions. A good way to develop skills in both sketching and lettering is to practice.

Schematic drawings differ from sketches. Although schematic drawings show the relationships of objects, usually they are not proportional. Such drawings usually are provided by manufacturers of equipment and parts. Hydraulic equipment, for example, often has a schematic to illustrate how the circuit is planned. A service manual usually has schematic drawings. One drawing, for example, may illustrate how the hydraulic pump is connected to aid the tractor mechanic in repair.

Orthographic drawings present an exact replication of the three basic views of an object: front, right side, and top. The other views may be included if they are different from the basic views. An orthographic drawing provides the views as if one were actually standing at right angles to the view. In this way, every dimension can be seen and measured.

The draftsperson uses symbols to fully explain the drawing, and the reader must understand these to get a mental picture of the object being drawn. Although some agricultural workers may do an occasional orthographic drawing, it is more important that all workers know how to read and follow orthographic drawings.

Hand lettering is necessary to provide the dimensions and instructions to accompany the drawing. There are many styles, but the single-stroke commercial style should be mastered first, since it is the one most commonly used. Neat, legible plans show that the planners have organized their work and done it with care. Proper lettering won't make the plan better, but it certainly will make it more understandable.

THINKING IT THROUGH

1. List three common types of plans and describe how each might be used.
2. What equipment is necessary for sketching?
3. In a drawing what would (*a*) a solid line, (*b*) a broken line, and (*c*) a line of long and short dashes illustrate?
4. Using ruled paper, practice hand-lettering the alphabet in the style shown in Figure 4-10.

CHAPTER 5

READING A PLAN AND ORDERING MATERIALS

Signs and symbols play important parts in our lives. Scouts and explorers have to learn how to "read" trail signs, for example. Anyone who has watched football realizes how important it is to the understanding of the game to know what the game officials mean when they give certain signs with their arms. In baseball, the course of an inning and game often is determined by the signs the team manager flashes to the third-base coach and the coach passes on to players.

Punctuation is a system of language symbols. In speaking, a question can be indicated simply by changing the level, or inflection of the voice. But in writing, the question must be indicated with a symbol—the question mark.

A plan (drawing or sketch) uses signs and symbols to communicate ideas and instruction to someone else. The construction crew working on a farm building uses a series of plans to show everything from how the foundation should be placed to the location of electrical circuits. The beginning agricultural student may use only a single sheet of paper to draw a plan showing the dimensions of a sawhorse or shop stand.

However complicated or simple the plan, the work will be done more quickly and precisely if the workers know how to read the plan and follow its instructions.

CHAPTER GOALS

In this chapter your goals are:

- To read a sketch or drawing
- To select or determine the scale
- To use detail drawings
- To order necessary materials for a project

Reading a Plan

To understand a plan, it is necessary to know where to look for the basic information about the plan. It would be embarrassing and costly for a person to try working from a plan without first knowing what the plan is for. Imagine a farm worker setting out to build a large tool shed without reading the description of the plan being used and finding out only later that the plan was for a garage and not a tool shed.

It also is important to be able to convert a scale that may be in inches and frac-

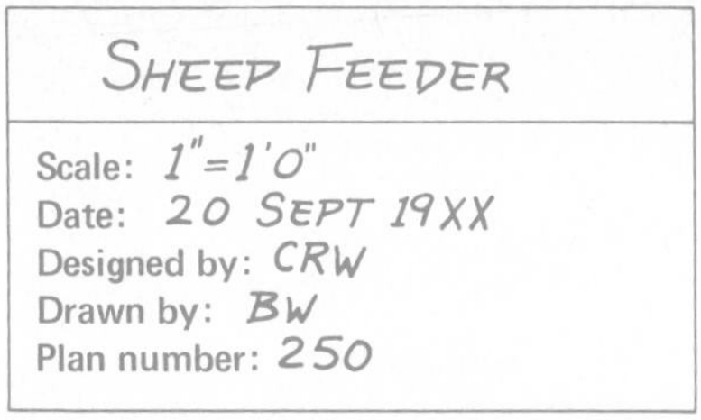

```
SHEEP FEEDER

Scale:  1"=1'0"
Date:   20 SEPT 19XX
Designed by: CRW
Drawn by:  BW
Plan number: 250
```

Figure 5-1. A typical title block as it would appear in the lower right-hand corner of a plan.

tions of inches into feet, or centimeters into meters.

Title Block

The *title block* on a plan is the first step in trying to find out what the plan is all about and what is important to know. The title block might be compared to the headline on an article in a newspaper or magazine. The title block is a small box usually placed in the lower right-hand corner of the drawing. It generally contains the kind of information that will help one get started using the plan (see Figure 5-1).

The name of the project should appear at the top of the block. In the example shown in Figure 5-1, the person picking up the plan knows immediately that the sketches and instructions all pertain to the construction of a sheep feeder. The next piece of information is the *scale*. This is vital. The plan cannot be put into effect—construction cannot begin—unless the worker knows what the scale is. The scale indicates the proportion of the dimensions as shown on the plan to the dimensions of the finished project. A ¾ scale, for example, means that ¾ in on the plan equals 1 ft [0.3 m] on the finished project. Other common scales are full, ½, ¼, and ⅛.

The date of the plan can be very important. For one thing, it tells whether this is the latest of several plans for the same project. It is possible that older plans for a sheep feeder exist that were not used and should not be used. The date also gives a point of reference for purchase orders and correspondence that have to do with the plan.

The title block also indicates who designed the project and who drew the plan. Again, this information can be important. If plans for the same project were submitted by different persons, the name in the title block would be the key reference in some instances.

Plans generally are numbered for purposes of reference and filing. If plans are filed by number, it makes it quite easy to get the plan desired. The number, of course, can also be important to the worker trying to select the right plan out of several for the same project.

Scales

A scale is used to show the proportions between the drawing and the actual object. In some cases the object is small enough that the drawing can show its exact size. This is called *full scale*. It means that any line shown on the drawing is exactly equal to the actual dimensions of the object. A tool-reconditioning template, for example, may be drawn to full scale because it is not a very large object (see Figure 5-2). In most cases, however, full-scale drawings are not possible because few objects are so small that they can be drawn full scale on paper of a reasonable working size.

Three Types of Scales.   The three types of scales commonly used in agricultural industry are the architect's scale, the engineer's scale, and the metric scale. The architect's scale is 12 in [30.5 cm] long and is used where dimensions are given in feet and inches. Plans of farm buildings and other carpentry projects usually give di-

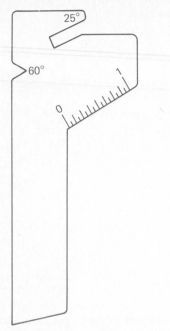

Figure 5-2.   A small object drawn to full scale.

mensions in feet and inches. Common architects' scales (see Table 5-1) show what portion of a foot on the plan equals a foot on the finished project. For example, the scale might use ½ in to represent 1 ft or ¾ in to represent 1 ft.

The engineer's scale also is 12 in long, but a 1 in division may represent 10 ft or 30 ft, and so on. The engineer's scale is used for drawing large areas such as a field, pond, or dam, or where dimensions are given in feet and tenths of a foot.

TABLE 5-1.   Common Architects' Scale

1½ scale	1½″ = 1′ 0″
Full scale	1″ = 1′ 0″
¾ scale	¾″ = 1′ 0″
½ scale	½″ = 1′ 0″
⅜ scale	⅜″ = 1′ 0″
¼ scale	¼″ = 1′ 0″
3/16 scale	3/16″ = 1′ 0″
⅛ scale	⅛″ = 1′ 0″

The metric scale is 30 cm long and is used to show proportions in metric measurements. The usual scales are 01, 02, 03, 05, 025, and 0125 (divisions per centimeter). The metric scale will become more common as the United States increases its use of metric measures.

The three considerations in selecting a scale are (1) the size of the original object, (2) the desired size of the drawing, and (3) the type of measurement being used. The scale also enables the worker to take the plans and convert them back to the exact dimensions of the original object.

Reading Orthographic Drawings

The value of these kinds of drawings is that they show measurements for all sides of an object. It is possible, for example, to show six different views of a sheep feeder by using an orthographic drawing. Figures 5-3 and 5-4 show the difference between a typical pictorial view and the presentation of six views with an orthographic drawing. The three basic views of an object are top, front, and side (sometimes called *end*). Most agricultural projects are drawn using just these fundamental views.

The top view may be called the *floor plan*. It shows the object as if viewed from above. The top view of the sheep feeder (Figure 5-5) gives a good idea of the floor area of the feeder. The dimensions given tell precisely the size of the feeder.

The front view would show the feeder

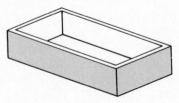

Figure 5-3.   A typical pictorial view of a box.

SKETCHING, DRAWING, AND PLAN READING

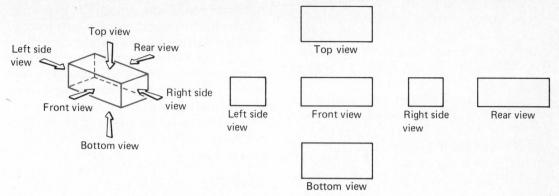

Figure 5-4. Six views and an orthographic drawing.

as viewed from the side which is designated as the "front." Generally, the front is the largest or most prominent portion of an object and the one having the most detail. This view also shows the shape, the height, and the length or width of an object.

The side (end) view of the feeder would show the shape and dimensions of one of its sides. If the opposite side differed markedly, then both a right and a left side must be developed and titled to correctly identify each.

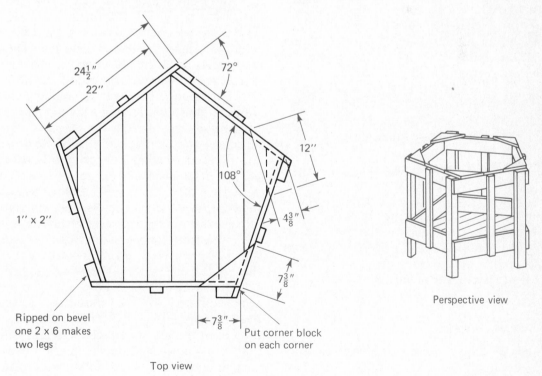

Top view

Figure 5-5. The top view of a sheep feeder.

Detail Drawings

Some projects, such as a landscape plan or a building, are so large that a major reduction in scale is necessary to fit the dimensions of the project on a single sheet of paper. When this happens, important details may be lost. For example, an aerial photograph of one square mile would not be too hard to understand if the picture were 16×20 in [40.6×50.8 cm]. But if the picture were reduced to 5×7 in [12.7×17.8 cm], then much of the detail would be lost.

If a worker is using an engineer's scale with 1 in [2.54 cm] on the aerial photograph equal to 660 ft [201 m], a portion of the project that measured less than 100 ft [30 m] would be proportionally so small on the plan that it would be nearly impossible to show properly. In such cases, therefore, small details may be left out of the major drawing. Special drawings may be prepared to show the small or detailed parts of the project. If one took an aerial photograph of a city but wanted more detail on just one block of the city, that portion of the photograph could be enlarged.

The detail drawing, then, takes a small part of the project and shows it in a different scale. The scale on the detail drawing, for example, might be full scale or ¾ scale.

Detail drawings usually have their own title box, even though they are related to larger drawings. In working with plans for a large project, it is a good idea to check the title box for a reference to detail drawings that may accompany the main drawing. A notation in the title box that the main drawing is only one of three sheets indicates that other drawings will follow.

Assembly Drawings

Many projects require a complex assembly procedure. Mechanics who work on equipment and vehicles often use assembly drawings to help them put together pieces of equipment and to take apart equipment for repair and replacement.

Ordinarily, an assembly drawing is accompanied by a *parts list*. The parts list enables the worker to identify the parts shown in the assembly drawing and helps in the ordering of replacement parts. Figure 5-6 shows an assembly drawing for a carburetor.

Ordering Materials

Plans are designed to help complete a project accurately and efficiently. The finished project should look like the plan and have the proper dimensions.

After reading the plan, it is necessary to get the materials needed to complete the project.

Developing a Cutting Bill of Materials

The plan, of course, is the guide to the materials that are needed. The plan must be read correctly and thoroughly if the materials are to be available to do the job.

Perhaps the best way to go about this process, particularly with a simple plan, is to mentally go through the steps of construction or assembly and then write down the materials needed for each step. A list of each item needed and its exact length is called a *cutting bill*. For example, if a loading chute were to be constructed, it would be necessary to figure out essential steps in the chute construction (see Figure 5-7).

For each step in the construction of the chute, the materials needed must be noted as well as the dimensions or other specifications for the materials. If the drawing is full scale, figuring the lengths of lumber or metal is relatively simple. If the plan uses another scale, it will be necessary to convert the plan's dimensions into actual dimensions.

SKETCHING, DRAWING, AND PLAN READING

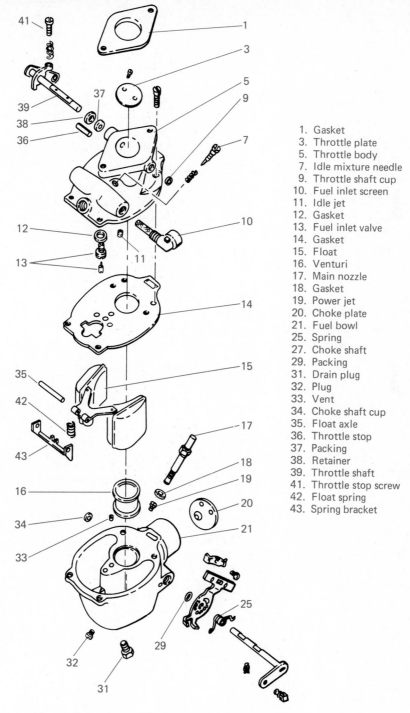

1. Gasket
3. Throttle plate
5. Throttle body
7. Idle mixture needle
9. Throttle shaft cup
10. Fuel inlet screen
11. Idle jet
12. Gasket
13. Fuel inlet valve
14. Gasket
15. Float
16. Venturi
17. Main nozzle
18. Gasket
19. Power jet
20. Choke plate
21. Fuel bowl
25. Spring
27. Choke shaft
29. Packing
31. Drain plug
32. Plug
33. Vent
34. Choke shaft cup
35. Float axle
36. Throttle stop
37. Packing
38. Retainer
39. Throttle shaft
41. Throttle stop screw
42. Float spring
43. Spring bracket

Figure 5-6.   Assembly drawing and parts list of a carburetor.

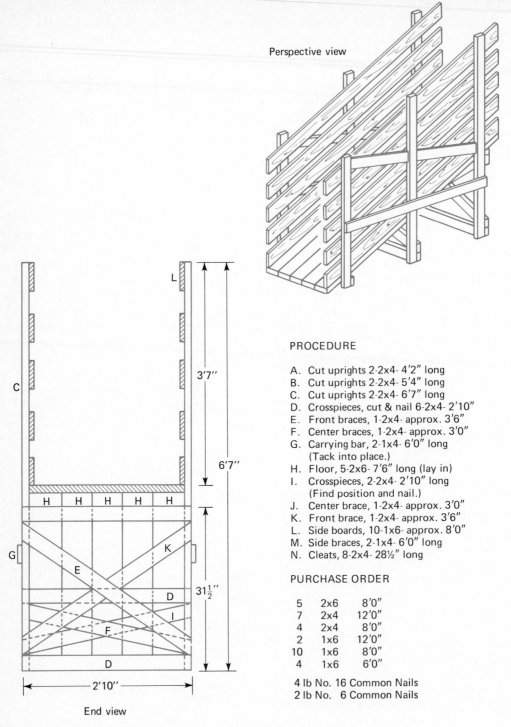

Perspective view

PROCEDURE

A. Cut uprights 2-2x4- 4'2" long
B. Cut uprights 2-2x4- 5'4" long
C. Cut uprights 2-2x4- 6'7" long
D. Crosspieces, cut & nail 6-2x4- 2'10"
E. Front braces, 1-2x4- approx. 3'6"
F. Center braces, 1-2x4- approx. 3'0"
G. Carrying bar, 2-1x4- 6'0" long
   (Tack into place.)
H. Floor, 5-2x6- 7'6" long (lay in)
I. Crosspieces, 2-2x4- 2'10" long
   (Find position and nail.)
J. Center brace, 1-2x4- approx. 3'0"
K. Front brace, 1-2x4- approx. 3'6"
L. Side boards, 10-1x6- approx. 8'0"
M. Side braces, 2-1x4- 6'0" long
N. Cleats, 8-2x4- 28½" long

PURCHASE ORDER

5	2x6	8'0"
7	2x4	12'0"
4	2x4	8'0"
2	1x6	12'0"
10	1x6	8'0"
4	1x6	6'0"

4 lb No. 16 Common Nails
2 lb No. 6 Common Nails

End view

Figure 5-7. Developing a cutting bill of materials for a completed plan.

SKETCHING, DRAWING, AND PLAN READING

TABLE 5-2.   Bill of Materials for Loading Chute

2	2 × 4	4′ 2″	Uprights
2	2 × 4	5′ 4″	Uprights
2	2 × 4	6′ 7″	Uprights
6	2 × 4	2′ 10″	Crossmember
1	2 × 4	3′ 6″	Front brace
1	2 × 4	3′ 0″	Center brace
2	1 × 4	6′ 0″	Carrying bar
5	2 × 6	7′ 6″	Floor
2	2 × 4	2′ 10″	Crossmember
1	2 × 4	3′ 0″	Center brace
1	2 × 4	3′ 6″	Front brace
10	1 × 6	8′ 0″	Sides
2	1 × 4	6′ 0″	Side brace
8	2 × 4	2′ 4½″	Cleats

While making note of the lumber or metal needed for the job, one must also identify the hardware needed. This may include nuts, bolts, screws, nails, hinges, hooks, or other parts.

A bill of materials like the one in Table 5-2 can be developed for the job.

Itemizing an Order Bill of Materials

The cutting bill of materials follows the plan and lists each item needed and its exact length. That kind of list, however, would be not very helpful to the lumber yard, because the lumber yard sells wood in standard lengths. Therefore, it is necessary to make up an order bill that is more useful to the places supplying the materials. The *order bill* is developed by combining materials and arranging them in a list, with the largest materials heading the list. The cutting bill of materials stated that five 2 × 6 pieces of lumber were needed and that each piece had to be 7 ft 6 in [2.3 m] long. But the lumber yard does not supply lumber in odd lengths. The order bill, therefore, lists five pieces that are each 8 ft [2.4 m] long. The workers will have to cut the extra 6 in [15.2 cm] off each piece when beginning the construction.

The order bill also combines the 28 pieces, which are 2 in [5.1 cm] thick and 4 in [10.2 cm] wide but of varying lengths, into seven 2 × 4 pieces that are each 12 ft [3.7 m] in length and four 2 × 4 pieces that are each 8 ft [2.4 m] in length. The construction worker will then cut the 28 pieces needed for the job out of these 11 pieces.

How to Read Plans and Order Materials: A Review

Plans are developed to aid the student or worker in constructing a project. These plans often use various symbols to reduce the amount of writing on the plan. These symbols as well as other important information usually are explained in the title box. Most plans are drawn to a scale where the dimensions shown on the plan are in some proportion to the actual dimensions. Some common architects' scales are full, ¾, ½, ¼, and ⅛.

Sometimes the important details of a plan for a very large project can be lost because of the scale used. When this happens, the details are drawn separately to a different scale.

When assembling a complex piece of equipment, workers often use an assembly drawing provided by the manufacturer. This drawing usually has a parts list as a reference.

Before a plan can be used, it is necessary to obtain the required materials and parts. A cutting bill of materials and parts list are developed to determine exactly what will be needed to complete the job. The cutting bill of materials also serves as a guide to construction workers. But an order bill of materials is prepared and given to the supplier of the materials. This order lists materials according to the standard dimensions handled by the supplier. By carefully combining materials, it is possible to reduce both cost and waste.

THINKING IT THROUGH

1. What is the purpose of the title box, and what information is usually found there?
2. Why do some drawings use an architect's scale while others use an engineer's scale?
3. When should a detail drawing be used?
4. What is the difference between a cutting bill of materials and an order bill of materials?
5. Why is it important to be able to determine the bill of materials from a plan?

CHAPTER 6

PROCEDURES FOR SKETCHING AND DRAWING

No matter what their career field of agricultural industry is, workers often need to figure out ways to do things and go ahead and draw a plan that will enable the project to be accomplished.

Many students in vocational agriculture develop livestock projects as a part of their supervised occupational experience program. As the project expands, there is a need for more livestock equipment. This may be purchased or it may be constructed as a part of the agricultural mechanics instruction during the school year. Many times valuable skills may be developed and, at the same time, the cost of the equipment may be reduced by constructing it in the shop.

Before a project is started, a plan should be developed to eliminate as many problems as possible. There are many good plans available for a variety of shop projects. However, if a plan is not available for the specific project, one must be developed. This chapter will show the steps necessary to draw a cattle feeder.

CHAPTER GOALS

In this chapter your goals are:

- To demonstrate an awareness of the value of well-developed orthographic drawings and describe how they can aid in the construction process
- To list the steps in getting ready to make the drawing
- To demonstrate the procedures used to draw the object
- To describe the steps taken to finish the job

Getting Ready

There are four important steps to follow before beginning an orthographic drawing. If the steps are followed, the drawing will be more accurate and more professional-looking.

Determining the Views

Most projects can be properly described in an orthographic drawing showing three views. To determine the views to be shown in the drawing, it is necessary first to make a rough sketch of the object, in this case a cattle feeder. By examining the sketch, it should be possible to decide what views are most important to show. In most cases,

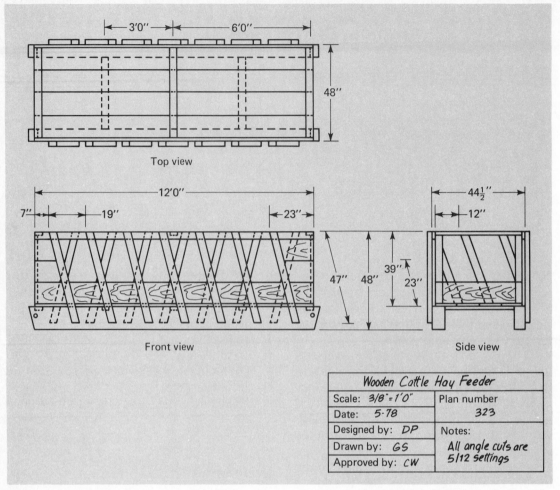

Top view

Front view

Side view

Wooden Cattle Hay Feeder	
Scale: 3/8"=1'0"	Plan number
Date: 5-78	323
Designed by: DP	Notes:
Drawn by: GS	All angle cuts are
Approved by: CW	5/12 settings

Figure 6-1. Orthographic drawing of a wooden cattle hay feeder.

the front and top views and one side view are sufficient. These views were selected for the cattle feeder. For some projects, of course, it may be desirable to also show a bottom view or another side view.

Generally the front view is selected as the basic view because it usually gives the longest dimensions and provides an overall impression of the shape of the project. The front view is placed to the left and below the middle of the paper. The top view is placed above the front view and on a verti-cal line with it (see Figure 6-1). The bottom view (not shown for the cattle feeder) is placed below the front view. The right side view is placed to the right of the front view, and the left side view (not shown for the cattle feeder) is placed to the left. The side views should be properly lined up horizontally with the front view (see Figure 6-1). If the rear view is shown (it is not shown for the cattle feeder), it is placed next to the right side view and on the same horizontal line as the front and side views.

SKETCHING, DRAWING, AND PLAN READING

It is helpful to make a rough drawing of the views in their proper positions before starting the final orthographic drawing.

Selecting the Scale

The second step before actually starting the final drawing is to select the proper scale. It is desirable to show the dimensions of the project as large as possible on the drawing. But the dimensions also have to fit on the paper being used. In the drawing for the cattle feeder, the scale selected is ⅜ in equals 1 ft [0.9 cm equals 0.3 m]. Using this scale, it was possible to easily place the three basic views on a sheet of paper 8½ × 11 in.

On the other hand, if the project were an 18-in [45.7-cm] electrode holder for the welding shop, it might be desirable to select an architect's scale and use the 3 in equals 1 ft [7.6 cm equals 0.3 m] scale. The object as drawn would then be 4.5 in [11.4 cm] tall.

Whatever scale is selected, it should be shown in the title block (see Figure 6-1).

Placing the Title Block

Before you start to draw, the third step is to locate the title block on the plan. Traditionally the title block is located in the lower right-hand corner of the drawing (see Figure 6-1). If large sheets are used for the drawing, the title block may be placed in the right-hand corner of an outside fold where it can be plainly seen. It is important that the title block can be seen without having to unfold the plan. In outlining the title block, enough space should be left to include all the necessary information.

Preparing the Drawing Paper

The final step in the "get ready" phase is to determine the center of the sheet and the base lines. Determine the halfway point on all four sides of the paper. Then draw a very fine line connecting the opposite points—top to bottom, side to side. Where the lines intersect will be the center of the paper. These lines will be erased after the object lines are established.

Drawing the Object

After you have finished the get-ready steps, it is time to start the orthographic drawing. The rough sketches of the views in the get-ready stage should have given an idea of what the finished drawing will eventually look like.

Laying Out the Dimensions

In the drawing step 1 is to arrange the various views and draw the principal dimensions of those views (use a light pencil line). The principal dimensions are those that give the overall outline of the object. Figure 6-2 shows the main dimensions.

Blocking In the Views

In step 2, other interior lines are added to the various views (see Figure 6-3). Some lines may need to be darkened with the pencil so that all lines are of equal blackness and width.

Putting In Dimension Lines and Lettering

The next step is to put in all dimension lines and measurements (see Figure 6-3). *Dimension lines* indicate the measurements, or distances between points. After the dimension lines are drawn, all necessary lettering should be done (following the instructions given in Chapter 5). The lettering includes putting in the numerals that designate measurements.

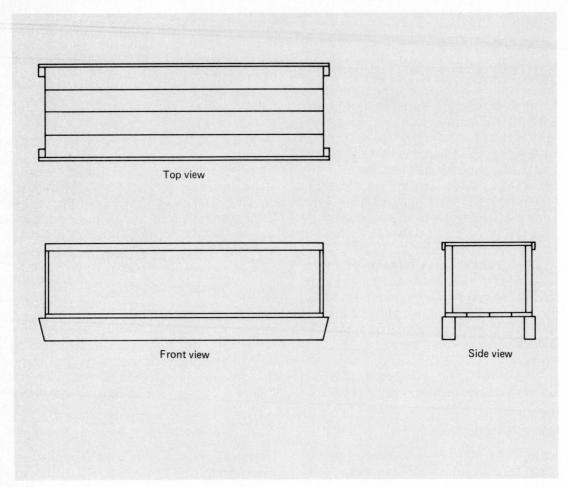

Top view

Front view

Side view

Figure 6-2.   Arranging the views and dimensions.

Drawing the Details

In some plans, it is necessary to show certain sections of the plan in more detail (see the section ''Scales''). The fourth step is to draw the center lines for that portion of the object to be shown in detail and then to draw that portion of the object. A new scale probably will have to be determined for the detail drawing. If on a separate page, the detail drawing should have its own title block.

Finishing the Job

The goal is to produce an orthographic drawing that will be as helpful as possible to those who have to use it. The plan should look finished—professional. Figure 6-4 shows the completed orthographic drawing of the cattle feeder. In the drawing of the cattle feeder notice that all lines have been drawn properly and that all lettering has been done neatly. It is important to

SKETCHING, DRAWING, AND PLAN READING

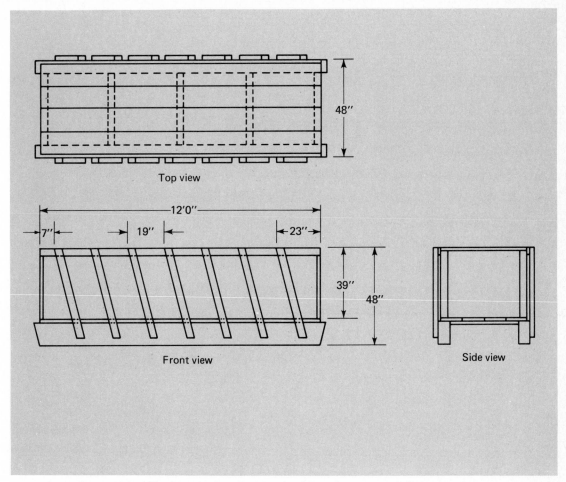

Figure 6-3. Blocking in the views and adding the principal dimension lines.

carefully review the drawing after completion to make sure everything is as it should be.

The planner must make sure that all necessary information is properly placed in the title block.

Remember, someone has to be able to read and follow the orthographic drawing. Mistakes on the drawing may result in mistakes in the finished object. That could prove very costly in many ways.

Procedures for Drawing: A Review

An orthographic drawing is the most common drawing used in agricultural industry. Although the sketch may be used exclusively for some very simple projects, many sketches are later turned into orthographic drawings. The orthographic drawing has two advantages: it is relatively easy to do, and it provides actual measurements be-

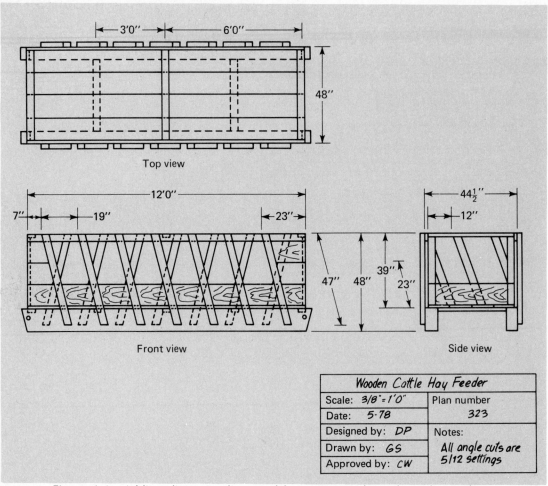

Figure 6-4. Adding dimension lines and lettering completes the orthographic drawing.

cause it is drawn according to a predetermined scale.

The four basic (get-ready) steps in making an orthographic drawing are as follows: (1) decide how many views should be illustrated (the three basic views being front, side, and top); (2) select the right scale so that the views of the object fill the paper being used but do not exceed it; (3) locate the title block on the sheet and do a preliminary sketch of the views to locate them properly; (4) determine the center of the drawing paper to help in placing the various views.

The final drawing should be done in three or four stages. The outer dimensions of the object should be drawn first, to give the object its proper shape. Then the views should be blocked in by adding the appropriate interior lines. The third step is to add the dimension lines that show measurements and to do all lettering. A fourth and optional step is to draw separate sections of the object that should be shown in greater detail. Usually a different scale is used for detail drawing.

When the drawing is completed, it should be carefully examined to make sure that all steps have been followed. The good drawing probably will be used by several persons. It is worth taking the time, therefore, to make sure the job is done right the first time.

THINKING IT THROUGH

1. What are the advantages of an orthographic drawing compared to a sketch?
2. List the four get-ready steps necessary to begin an orthographic drawing. Explain why each step is important.
3. What determines the scale used for a drawing?
4. What are the necessary steps in the actual drawing? Why should these steps be completed in order?
5. What should be done after the actual drawing is complete?

UNIT III
WOODWORKING AND BASIC CARPENTRY

COMPETENCIES

COMPETENCIES	PRODUCTION AGRICULTURE — Small-animal producer	AGRICULTURAL SUPPLIES/SERVICES — Farm and garden center salesperson	AGRICULTURAL MECHANICS — Farm building construction worker	AGRICULTURAL PRODUCTS, PROCESSING, AND MARKETING — Fruit and vegetable shipping worker	HORTICULTURE — Groundskeeper	FORESTRY — Pulpwood worker	RENEWABLE NATURAL RESOURCES — Game warden
Select building materials	Very Important	Very Important	Very Important	Important	Important	Very Important	Not Important
Calculate board feet	Very Important	Very Important	Very Important	Important	Important	Very Important	Not Important
Determine quantities and costs of materials	Very Important	Very Important	Very Important	Important	Very Important	Very Important	Not Important
Accurately measure materials	Very Important	Very Important	Very Important	Very Important	Very Important	Not Important	Very Important
Use hand tools safely	Very Important	Very Important	Very Important	Very Important	Very Important	Not Important	Very Important
Identify the major parts of power tools	Very Important	Very Important	Very Important	Very Important	Important	Very Important	Important
Follow safety procedures while operating power tools	Very Important	Very Important	Very Important	Very Important	Important	Very Important	Important

 Very Important

 Important

 Not Important

Just as sketching and drawing skills are needed by many workers in the various agricultural career fields, so woodworking and basic carpentry skills are required by many workers.

For example, a number of workers in production agriculture, agricultural supplies/services, agricultural mechanics, agricultural products, processing, and marketing, and the other fields should know how to use hand and power woodworking tools safely and efficiently. When a number of jobs require common skills, these jobs are said to be in a *cluster*. Most of the time it is the mechanical skills discussed in this book that are common to job clusters.

The person who masters woodworking and basic carpentry skills, therefore, is not limited to only a few jobs in a few agricultural career fields. The skills will come in handy in many jobs in all career fields.

The farm worker may repair older buildings and construct new ones; the ranch worker may build a corral or a fence; the nursery worker may construct benches for a greenhouse; the person in charge of parts at the farm equipment dealership may put up new shelving and bins for small parts; and so on.

Upon completion of this unit, you should know how to select and use building materials and determine the amount of materials needed for a job. You also should know how to measure materials and figure their cost.

Finally, at the end of this unit you should know how to use hand and power woodworking tools properly and safely.

CHAPTER 7

SELECTING
BUILDING MATERIALS

Woodworking and carpentry skills are used to build or repair a number of items made entirely or partly of wood. These items may range from a machinery shed for farm equipment to a workbench in the shop of a mechanic who services agricultural equipment.

Whatever the project, initial planning is important. The worker must first determine the kinds and quantities of materials that will be needed.

One must be able to select the best lumber for the job and know how to determine board feet. It is also important to determine what fasteners and other hardware will be needed, such as nails, screws, paint, and preservative.

CHAPTER GOALS

In this chapter your goals are:

- To select the correct grade or standard of lumber for the job to be accomplished
- To calculate board feet
- To determine the quantity and total cost of materials needed
- To identify nails and screws commonly used in wood construction
- To identify and select glues for specific jobs

- To select paints, stains, and preservatives for specific jobs

Selecting Grades and Standards of Lumber

There are two basic kinds of wood: *hardwood* and *softwood*. These terms refer to the trees from which the wood comes rather than the actual hardness or softness of the wood. Hardwood comes from deciduous trees. *Deciduous* trees are those that shed their leaves each season. Examples are oak and walnut. Softwood comes from *coniferous* trees—those that do not shed their leaves at one time. Examples of coniferous trees are pine and cedar. A more complete listing of hardwoods and softwoods can be found in Table 7-1.

The most common hardwoods used in agricultural construction are red and white oak, walnut, hickory, and ash. Common softwoods are pine, fir, redwood, and cedar.

Wood that is dressed, or processed, for use in building is called *lumber*. There are two basic kinds of lumber: commercial (also called *graded*) and ungraded lumber. *Commercial lumber* is wood that has been dressed and graded by a large saw mill.

TABLE 7-1.   Kinds of Wood

SOFTWOODS	HARDWOODS
Cedar, Inland Red	Ash, White
Cedar, Western Red	Birch, Yellow
Cypress	Cottonwood
Fir, Douglas	Elm, Rock
Fir, White	Hickory, True
Hemlock, Western	Maple, Hard
Larch, Western	Oak, Red and White
Pine, Western White	Walnut
Pine, Lodge Pole	
Pine, Ponderosa	
Pine, Southern Yellow	
Pine, Sugar	
Redwood	
Spruce, Englemann	
Spruce, Sitka	
Tamarack	

Usually it has been planed and cut to standard lengths, widths, and thicknesses. *Ungraded lumber* is wood that has been sawed at small local mills. The wood may not be in standard lengths, and generally it has not been planed or sanded. Ungraded lumber is best for some jobs where quality finish is not needed and costs must be minimized. Ungraded lumber might be used, for example, for small structures such as animal feeders, or corrals and fences.

Commercial lumber is further classified as boards, dimension lumber, and timbers (see Table 7-2). *Boards* are lumber that is less than 2 in [51 mm] thick, but more than 2 in wide. *Dimension lumber* is between 2 and 5 in [51 and 127 mm] thick and more than 2 in wide. *Timbers* are at least 5 in thick and wide.

Calculating Board Feet

Lumber is ordered and paid for in board feet. A *board foot* is lumber that is 1 in [2.5 cm] thick, 1 ft wide, and 1 ft [30.5 cm] long. To determine board feet, multiply the number of pieces times the thickness in inches times the width in inches times the length in feet. Then divide by 12 in. (If the length is given in inches, divide by 144.) If five pieces of lumber are needed and each is 2 × 6 in [5.1 × 15.2 cm] and 12 ft [3.7 m] long, then the board feet are figured as follows: 5 × 2 × 6 × 12 ÷ 12. The correct number of board feet is 60. The cost of the lumber is 60 board feet multiplied by the price per board foot.

Commercial lumber is never full width or thickness. That is, the common 2 × 4 is not actually 2 in × 4 in because the width and thickness of the board are reduced by sawing, surfacing, and finishing operations at the sawmill. The dressed size would be closer to 1½ × 3½ in. However, in the example just given, the nominal width and thickness of the 2 × 4 would be used in computing board feet. Both the nominal and dressed sizes of lumber are used in Table 7-3.

Plywood

Plywood is a special kind of lumber. It is made by gluing together three or more thin sheets of lumber (plies). The grain of each sheet—the *ply*—runs in the opposite direction of the grain of the sheet on top of and below it (see Figure 7-1).

There is plywood for interior and exterior use. Exterior plywood is better able

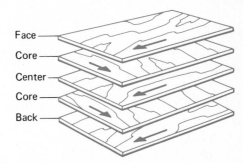

Figure 7-1.   Composition and grain direction of a sheet of plywood.

TABLE 7-2. Grades of Lumber.

CLASSIFICATION	GRADES	DESCRIPTION OR USE
DIMENSION 2″ to 5″ thick 2″ or more wide	STRUCTURAL LIGHT FRAMING (2″-4″ thick, 2″-4″ wide)	
	Select structural	• Use where high strength and stiffness and good appearance are needed.
	No. 1	• Use about the same as SEL STR, a little lower in quality.
	No. 2	• Recommended for most general construction uses.
	No. 3	• Use for general construction where appearance is not a factor.
	LIGHT FRAMING (2″-4″ thick, 2″-4″ wide)	• Provides good appearance where high strength and high appearance are not needed.
	Construction	• Recommended and widely used for general framing purposes.
	Standard	• About same uses as CONSTRUCTION but a little lower in quality.
	Utility	• Used for studding, blocking, plates, etc. where economy and good strength are desired.
	Economy	• Suitable for crating, bracing and temporary construction.
	STUDS (2″-4″ thick, 2″-4″ wide) Stud	• Only one grade; suitable for all stud uses.
	Structural Joists & Planks (2″-4″ thick, 6″ & wider)	
	Select Structural	• Use where high strength and stiffness and good appearance are needed.
	No. 1	• Use about same as SEL STR; a little lower quality.
	No. 2	• Recommended for most general construction use.
	No. 3	• For use in general construction where appearance is not a factor.
	APPEARANCE FRAMING (2″-4″ thick, 2″ & wider) A	• Use exposed in housing and light construction for high strength and finest appearance.
TIMBERS 5″ or more in least dimension	Select Structural	• Use where superior strength and good appearance are needed.
	No. 1	• Similar uses to SEL STR; a little lower in quality.

TABLE 7-2. Grades of Lumber. (Continued)

CLASSIFICATION	GRADES	DESCRIPTION OR USE
TIMBERS 5″ or more in least dimension	No. 2 No. 3	• Recommended for general construction. • Use for rough general construction.
BOARDS up to 1½″ thick, 2″ or more wide	 SEL MER or 1 CONST 2 STD 3 UTIL 4 ECON 5	• Graded for suitability for use in construction. • Use in housing and light construction for exposed paneling, shelving, etc. • Use for subfloors, roof sheathing, etc. • Used about the same as #2 but a little lower in quality. • Combines usefulness and low cost for general construction purposes. • Use for low grade sheathing, crating and bracing.

Source: *Structures and Environment Handbook*, Midwest Plan Service, seventh edition.

TABLE 7-3. Nominal and Minimum-Dressed Sizes of Boards, Dimension, and Timbers.

(The thicknesses apply to all widths and all widths to all thicknesses.)

ITEM	THICKNESSES (in inches) Nominal	THICKNESSES (in inches) Dressed Dry	FACE WIDTHS (in inches) Nominal	FACE WIDTHS (in inches) Dressed Dry	ITEM	THICKNESSES (in inches) Nominal	THICKNESSES (in inches) Dressed Dry	FACE WIDTHS (in inches) Nominal	FACE WIDTHS (in inches) Dressed Dry
Boards	1″	¾″	2″	1½″	Dimension	2	1½	2	1½
	1¼	1	3	2½		2½	2	3	2½
	1½	1¼	4	3½		3	2½	4	3½
			5	4½		3½	3	5	4½
			6	5½		4	3½	6	5½
			7	6½		4½	4	8	7¼
			8	7¼				10	9¼
			9	8¼				12	11¼
			10	9¼					
			11	10¼					
			12	11¼	Timbers	5 and	½ off	5 and	½ off
			14	13¼	(Dressed	thicker		wider	
			16	15¼	green)				

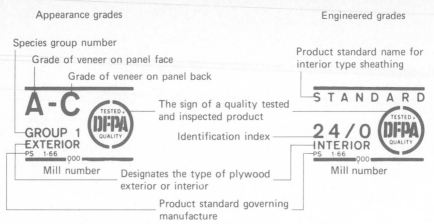

Figure 7-2. Various grades stamped on plywood.

to stand up to weather conditions because it is made with waterproof glue. Plywood is graded as follows: A or N, highest quality with no knots or patches; B, some small round knots and some patches or small plugs; C, more small knots and patches and some knotholes; D, more and larger knots and knotholes. The C grade is the lowest allowed for exterior plywood. Each side of the sheet is graded separately. Thus, a sheet may be graded A on one side and D on the other. Both grades generally are stamped on the back of the panel (see Figure 7-2).

Plywood has a wide variety of uses. A low-grade (C) plywood can be used for outside walls of a small shed, and a high-grade (A) plywood can be used as paneling for an agribusiness office. Often plywood is used as sheeting in animal shelters; as forms for foundations or concrete walls of shops and farm buildings; and as rough flooring under carpet, tile, or finished hardwood in offices and other buildings.

Selecting Nails and Screws

Different Kinds of Nails

Nails can be used to secure or fasten two or more pieces of lumber. Nails are manufac-tured from mild steel, aluminum, or other metals. Some nails are coated with resin, cement, or other substances to prevent them from pulling out easily. Other nails are specially treated to make the nail sur-face rough so that it will resist withdrawal.

There are many different kinds of nails (see Figure 7-3). Some are designed to be used just on roof shingles; others are made only for plasterboard or concrete. Some have large heads, and some, such as the finishing nail, have almost no head at all. Nails, such as a box nail, have a long, sharp point that prevents splitting. The common nail has a point shaped like a diamond.

In England, the common nail was des-ignated as a "penny," such as a 6-penny or a 10-penny. Some persons believe the labels came about because the nails were sold for sixpence and tenpence per hundred. The penny labels have con-tinued, but today it is written as a 6d nail rather than a 6-penny nail. The letter d stands for penny. To find out how long a nail is, divide the "d number" by 4 and add ½. The 6d nail, for example, is 2 in [51 mm] long.

As might be expected, the common nail is used most often. Common nails are kept on hand for constructing and repairing buildings, fencing, and a variety of wood-

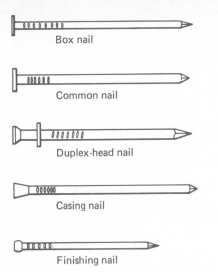

Box nail

Common nail

Duplex-head nail

Casing nail

Finishing nail

Figure 7-3. Examples of common nails used in agricultural construction.

TABLE 7-4.   Woodscrew Data

SHANK DIAMETER	SCREW LENGTHS*	SHANK HOLE SIZE	PILOT HOLE SIZE
0	$3/16$-$1/4$	$1/16$	$1/64$
1	$3/16$-$1/4$	$5/64$	$1/32$
2	$3/16$-$5/8$	$3/32$	$1/32$
3	$3/16$-$5/8$	$7/64$	$3/64$
4	$3/16$-$1/4$	$7/64$	$3/64$
5	$3/8$ -$1/4$	$1/8$	$1/16$
6	$3/8$ -$1/2$	$9/64$	$1/16$
7	$1/2$ -$1/2$	$5/32$	$1/16$
8	$1/2$ -$2/2$	$11/64$	$5/64$
9	$5/8$ -$2/2$	$3/16$	$5/64$
10	$5/8$ -$3/2$	$3/16$	$3/32$
12	$5/8$ -$3/2$	$15/64$	$3/32$
14	$5/8$ -4	$7/32$	$7/64$
16	$5/8$ -4	$1/4$	$7/64$

* Common screw sizes in inches are: $3/16$, $1/4$, $3/8$, $1/2$, $5/8$, $3/4$, $7/8$, 1, $1 1/4$, $1 1/2$, $1 3/4$, 2, $2 1/4$, $2 1/2$, $2 3/4$, 3, $3 1/2$, and 4.

working projects. The box nail is slightly smaller in diameter; often it is used when the thicker common nail might split the wood. The box nail is good for nailing siding on buildings.

Finishing nails are used primarily for interior construction and trim work. Because they have no head, they are easily driven below the surface of the wood. Roofing, shingling, and installing plasterboard all require other special nails.

Nails are more secure when driven into wood perpendicular to the grain (see Figure 7-4 for an example of wood grain). When the nail is driven in at a slight angle, the bond is even stronger.

Screws Secure Wood Better

Screws hold wood together much more securely than nails. Like nails, they come in different sizes and shapes for a variety of uses (see Table 7-4). The three common types of screws are the flathead, roundhead, and ovalhead (see Figure 7-5). There are two common variations among the heads. One is the slotted head, and the other is the Phillips (cross) head.

To make it easier to drive the screw with the screwdriver, first a hole should be drilled into which the screw will be inserted. Generally, a small, or pilot, hole is drilled first, with a drill bit considerably smaller than the diameter of the screw to be used. A second hole is drilled that is about the same diameter as the screw. Pilot drill bits are available that perform both functions at the same time (see Figure 7-6).

Countersinking means to set the head of a flathead screw even with or slightly below the surface of the wood (see Figure 7-7).

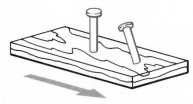

Figure 7-4. Driving a nail at a slight angle increases the bond of the joint.

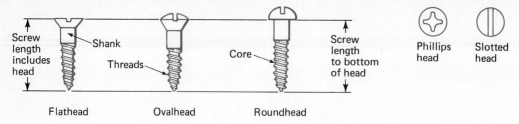

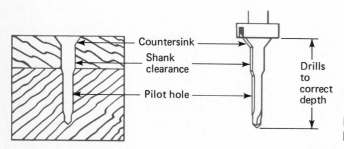

Figure 7-5. Three common wood screws.

Figure 7-6. A pilot bit can be used to set wood screws.

Selecting Glues and Paints

Glue Is Also a Fastener

Another way to fasten wood is with glue. As a general rule, it is easier to glue softwood than hardwood.

Before using glue, make sure that the wood is clean. Dust, oil, paint, or other foreign matter on the wood will prevent a secure bond. Not all glues are the same. As with most materials, different glues are de-signed for specific jobs. For example, some glues are more weatherproof than others. It would be unwise to use a glue that isn't weatherproof on an outside project. Consult Table 7-5 as a guide to the selection of common glues.

Glues take different times to set. The manufacturer's recommendations should be followed. To ensure the bond, the pieces being glued should be clamped or temporarily nailed. In some cases, it may be possible to lay the pieces flat and place weights on top while the glue sets. In some construction projects, it is common to use both nails or screws and glue.

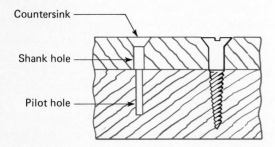

Figure 7-7. Countersinking sets the head of a flat head screw slightly below or even with the surface of the wood.

Coatings Preserve Wood

Most projects constructed of wood, whether they are a livestock shelter or a shop bench, require some kind of coating to preserve as well as decorate the wood.

Paints are of two basic types: *water-base* and *oil-base*. A paint with an oil base tends to penetrate better and lasts longer.

TABLE 7-5. Wood Glues

PROPERTY	SYNTHETIC RESIN GLUES					PROTEIN GLUES	
	Resor-cinol	Urea	Poly-vinyl	Epoxy	Contact Adhe-sives	Casein	Animal
Needs mixing	X	X		X		X	X
Crack-filling			X	X	X	X	
Applied hot							X
Applied cold	X	X	X	X	X	X	
Colorless glue line		X	X		X	X	X
Dark colored glue line	X						
Tends to stain certain woods						X	
Pressed at 70°	X	X	X	X	X	X	X
Over 8-hour working life		X	X				X
Low moisture resistance			X				X
Medium moisture resistance			X		X	X	
Good to high moisture resistance	X	X		X			
Low temperature resistance		X	X				
High temperature resistance	X			X	X	X	X
For structural gluing	X	X				X	
For exterior uses[1]	X			3			
For interior uses[2]		X	X	3	4	5	X

[1] Exterior uses include outdoor furniture, boats, and recreational equipment.

[2] Interior uses include furniture, cabinets, framing, and other shopwork that will be used in a moderately dry atmosphere.

[3] Excellent for bonding metal, plastics, and cloth to wood. No practical advantage on wood-to-wood gluing over resorcinol resin except it is a good joint filler.

[4] Covering counters and cabinets with leather, linoleum, and plastic laminates.

[5] Mold-resistant types are highly water resistant, though not waterproof.

Source: *Structure and Environment Handbook*, Midwest Plan Service, seventh edition.

But turpentine or other solvents must be used to clean the equipment after an oil-base paint is employed. When water-base paint is used, brushes and rollers can be cleaned with soap and water. Also, a water-base paint is usually lower in price.

Most new wood requires a *prime coat*. This is a coat of special materials that seal the surface and act as a bonding agent for the finish coat. Priming paint can be purchased at the same place and at the same time as paint for the finish coat.

Before you paint, the surface should be cleaned of dust, grease, or other materials that might resist the paint. The wood also must be dry.

Stains may be purchased with either a water or oil base. Stains are used mainly for interior trim. The directions on the label should be followed.

In some cases, it may be desirable to preserve the wood but not paint or stain it. For example, the farm worker may want to prevent fence posts from rotting but does not want to paint them. *Creosote* or *penta* can be used for this purpose. Both can be brushed on. However, it is usually better to buy lumber that already has been treated with creosote or penta. In such instances, the preservatives have been forced into the wood under heat and pressure.

Selecting Building Materials: A Review

The agricultural worker who has sketched or drawn plans for a construction project, no matter how large or small, has to be able to select the correct building materials if the project is going to be a success.

One might compare the agricultural worker with the chef who must select the right ingredients in order to make a recipe turn into something that is pleasing both to eat and to look at. Of course, the selection process for the pastry chef is quite different from that for the salad chef. Whipped cream would not go well in a tossed salad, and avocados would not be suitable for cherries jubilee.

It is also true that the agricultural worker planning to build a cattle feeder should know how to select the lumber and other materials that are correct for the weathering and use the feeder will get. Therefore, oak or cypress would be better than untreated fir or pine. The worker building a barn for livestock has to make sure waterproof glues are used. If non-waterproof glues are used, the moisture from the livestock will cause joint failure.

Lumber is grouped into hardwoods and softwoods. Also, it may come from a commercial sawmill or a local mill. If it comes from a small, local mill, the lumber may be ungraded. Lumber is further classified as boards, dimension lumber, and timbers.

Lumber is measured and generally sold in board feet. Board feet are determined by multiplying the width times the thickness times the length in feet and dividing by 12. A 2×6 in board that is 10 ft long contains 10 board feet. Costs often are based on dollars per 100 board feet. If the 2×6 were priced at $29 per 100 board feet, then 10 board feet would cost $2.90.

Plywood is a common building material in agricultural industry that is used for both outside and inside projects. The plywood is graded, with the grades of A or N indicating highest quality and D indicating that there are large knots or knotholes in the wood.

Different kinds of fasteners are used to hold pieces of lumber together. These fasteners include nails, screws, and glues. While screws are generally better fasteners than nails, some nails resist withdrawal better than others. Glue is widely used in agricultural structures and offers more strength than either screws or nails. A disadvantage of glue is that it may take more time to set and bond.

Almost without exception, the performance and life of a wood structure will be improved by using a paint, stain, or preservative. Creosote and penta are used to prevent rot. Stains and paints are used to improve appearance and to make the wood resistant to moisture. Paints and stains are available with either an oil or a water base. The oil-base product penetrates the wood better, but brushes and rollers are easier to clean after using a water-base paint or stain.

THINKING IT THROUGH

1. What are the common kinds of wood used in agricultural industry? Give examples of where each kind of wood could be used.
2. What is the difference between hardwoods and softwoods?
3. What are the grades of lumber used in agricultural construction?
4. Calculate the total board feet in the following bill of materials: (a) five 2 × 6 in × 6 ft; (b) three 1 × 2 in × 10 ft; (c) three 2 × 10 in × 6 ft; (d) three 1 × 8 in × 15 ft
5. Which lumber in the preceding problem is not commonly stocked in a lumber yard?
6. What factors should be considered in deciding whether to use nails or screws?
7. List the three most commonly used nails and describe their uses.
8. What are the advantages and disadvantages of using glue?
9. What are the advantages and disadvantages of using an oil-base versus a water-base paint?

CHAPTER 8

USING HAND WOODWORKING TOOLS

The agricultural worker who has made plans to construct something using wood and has selected the building materials necessary to do the job must know how to use woodworking tools if the job is to be completed properly.

Many workers in the various agricultural career fields need to know how to use woodworking tools. They may be constructing a new building or just making shelves for the storage of shop tools. The workers constructing the building probably will use a number of tools, while the person building the shelves may only use a few. In both cases, however, knowing how to use the tools correctly may make the difference between doing the job well and doing it poorly.

CHAPTER GOALS

In this chapter your goals are:

- To accurately measure materials and openings
- To identify common hand woodworking tools
- To select the correct tools for each job
- To use tools properly and safely

Using Measuring and Marking Tools

There is nothing quite so frustrating as cutting out pieces of lumber and then finding out they don't fit the intended opening. The problem usually is that the worker measured either the opening or the lumber (or both) incorrectly. To prevent this, it is important to learn how to use measuring and marking tools.

Common measuring tools are the framing square, try square, combination square, folding rule, and steel tape (see Figure 8-1). Most measuring tools today provide scales in both U.S. Customary and metric systems. The metric system is used in most of the world today other than the United States. However, the United States has begun a gradual conversion from the U.S. Customary System to the metric system.

The U.S. Customary System is divided into yards, feet, inches, and fractions of an inch. Usually, the U.S. Customary scale on woodworking tools is not reduced lower than $1/16$ in. The metric scale is divided into meters, centimeters, and millimeters. Since 1 mm is equal to approximately $4/100$ in, more accurate measurements are possible in the metric sys-

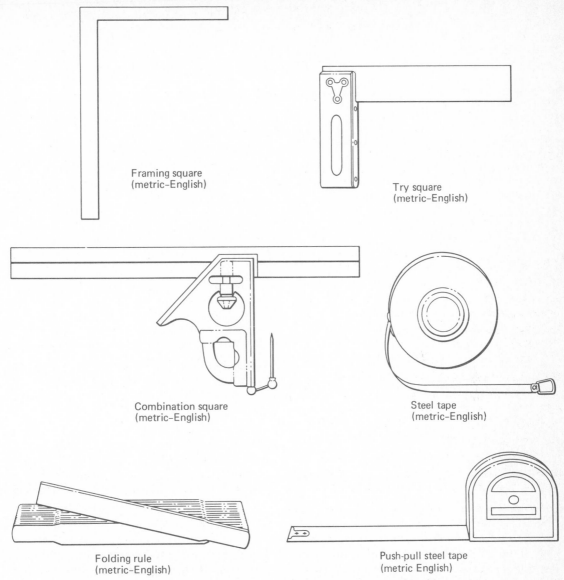

Framing square
(metric–English)

Try square
(metric–English)

Combination square
(metric–English)

Steel tape
(metric–English)

Folding rule
(metric–English)

Push-pull steel tape
(metric English)

Figure 8-1. Common measuring tools used in woodworking.

tem. Many times, the accuracy of the measurement depends largely on the skill and care taken by the worker.

Framing Square

The English framing square is in the shape of a large letter L. The longer part of the square (24 in) is called the *body*. The shorter part (16 in) is the *tongue*. The body is 2 in wide, and the tongue is 1½ in wide (see Figure 8-2).

There are measurement scales on both the face (front) and the back side of the square. The *face* is the side nearest a person holding the tongue of the square in the

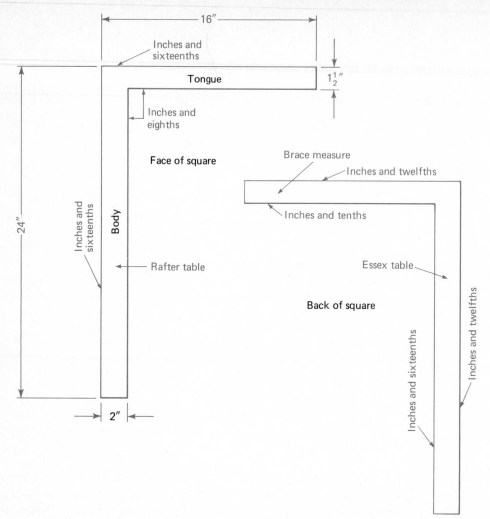

Figure 8-2. The measurement scales found on the face and back of a framing square.

right hand and the body in the left hand. On the face of most squares, both outside edges are scaled in inches and sixteenths of an inch. The inside edges are scaled in inches and eighths of an inch. On the back, both outside edges are scaled in inches and twelfths of an inch. The inside edge of the body is scaled in inches and sixteenths of an inch; the inside edge of the tongue in inches and tenths of an inch. This is done to allow for accurate measurement of fractions.

The framing square can be used to determine if the end of a board is square, that is, at a 90° angle to the side of the board. To make this determination, fit the square over the two edges of the board as shown in Figure 8-3. If the square does not fit snugly along the inside edges of blade and tongue, then the board is not square.

The framing square also can be used to square a board for sawing. By placing the square as shown in Figure 8-3, a straight line can be drawn on the board at a perfect

WOODWORKING AND BASIC CARPENTRY

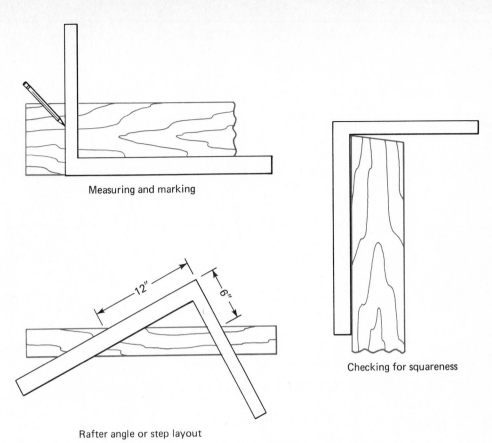

Measuring and marking

Checking for squareness

Rafter angle or step layout

Figure 8-3. Common uses of the framing square.

90° angle (square) to the edge of the board. When the saw follows this line, the resulting cut should be a square cut.

Other uses of the framing square are to mark lumber for building rafters, doing stair layout, marking angles, and figuring board feet of lumber. These uses will be discussed in later courses. The angle and grade chart (Figure 8-4) shows another way to use the square.

Try Square

The primary use of the *try square* is to check for the squareness of lumber and products made from wood. For example, a try square might be used to check whether the ends of a board are square with the sides. The try square also can be used to determine the squareness of a plane iron. Figure 8-5 shows some of the common uses of the try square.

The try square also is shaped like the letter L, but it is smaller than the framing square. Only the blade of the try square has a measuring scale on it. The short part of the square is called the *head*. To protect it from being damaged and to ensure accurate measurements, the head should be made of metal.

Combination Square

The *combination square* can be used for the same things as the framing and try squares. But because it has a movable

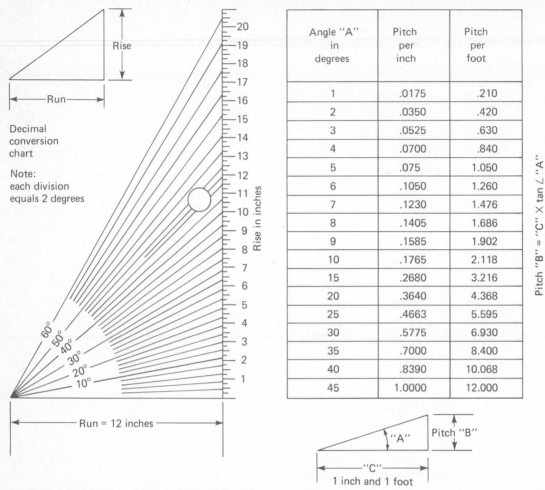

Angle "A" in degrees	Pitch per inch	Pitch per foot
1	.0175	.210
2	.0350	.420
3	.0525	.630
4	.0700	.840
5	.075	1.050
6	.1050	1.260
7	.1230	1.476
8	.1405	1.686
9	.1585	1.902
10	.1765	2.118
15	.2680	3.216
20	.3640	4.368
25	.4663	5.595
30	.5775	6.930
35	.7000	8.400
40	.8390	10.068
45	1.0000	12.000

Figure 8-4. The angle and grade chart converts degrees in angles to pitch per inch and pitch per foot.

head, its uses are more varied. It can be used as a straightedge, depth gauge, metric gauge, and marking gauge. Also, it can be used to lay out 45° angles (see Figure 8-6).

For example, suppose the worker wanted to mark a board 2 in [51 mm] in from the edge. The blade of the combination square should be positioned so that 2 in extends out from the head. Then, the head is held flush against the edge of the board. The thumbscrew on the head is tightened. Finally, the head is slid along

the edge while a pencil is used to mark along the end of the blade.

In another instance, the worker might want to measure the depth of a notch cut out of a board. The end of the blade should be placed into the notch. Then, the head is moved until it is flush with the outer edge of the board. Now, when the square is taken away, the depth will be recorded where the head was positioned. Other boards can easily be marked for a similar notch.

WOODWORKING AND BASIC CARPENTRY

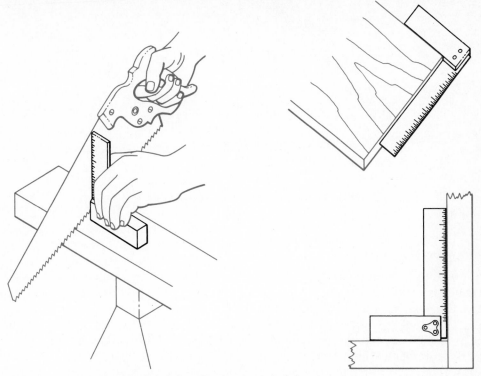

Figure 8-5.   Common uses of the try square.

Some combination squares include a leveling bubble built into the head. Depending on how the square is positioned, the bubble can be used to determine a plumb line or whether a board or other surface is level.

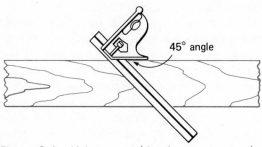

Figure 8-6.   Using a combination square to determine a 45° angle.

Figure 8-7 shows how a square or straightedge can be used to divide a board into equal pieces.

Sliding Bevel Square

The *sliding bevel square* is made of metal with a rosewood head. Like the combination square, it comes with a movable head. The difference, however, is that the head of the sliding bevel square can be pivoted in such a way as to create angles with the blade (see Figure 8-8). The sliding bevel square is used most often to transfer existing angles or to make a number of duplicate angles.

There are several ways to determine the angle wanted. A protractor, drafting

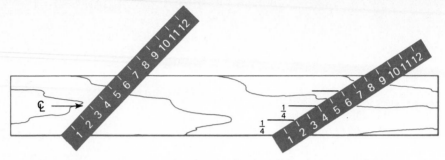

Figure 8-7. Using a straightedge to divide a board into equal parts.

triangle, or framing square may be used (see Figure 8-9).

Tape Measures and Folding Rules

The steel tape that retracts into a carrying case is the most convenient, practical, and often used measuring tool in agricultural industry. The steel tape can be used very efficiently to measure inside dimensions, such as the number of feet between two posts in a building. In using a steel tape with carrying case, it is important to note how many inches should be added on to the reading to take into account the portion of the tape inside the case.

Steel tapes can be purchased that have fractions of an inch as well as metrics. Some tapes have tables on them to permit conversion from the U.S. Customary System to metric systems, or vice versa. The tapes also may include formulas and tables on the blade for figuring volume, distance, or rafter length for example.

The steel tape has a tip that can hook over one end of a piece of material while the tape is moved down the length to make the measurement.

Folding rules are not as commonly used as steel tapes. The folding rule comes with a variety of measuring scales printed on it, including metrics. The better folding rules have a 6-in [15.2-cm] sliding rule at one end to help in taking inside measurements.

Both the steel tape and folding rule are easily damaged. Therefore, care should be taken with them. The steel tape should always be returned to its case after use, and the folding rule should be refolded.

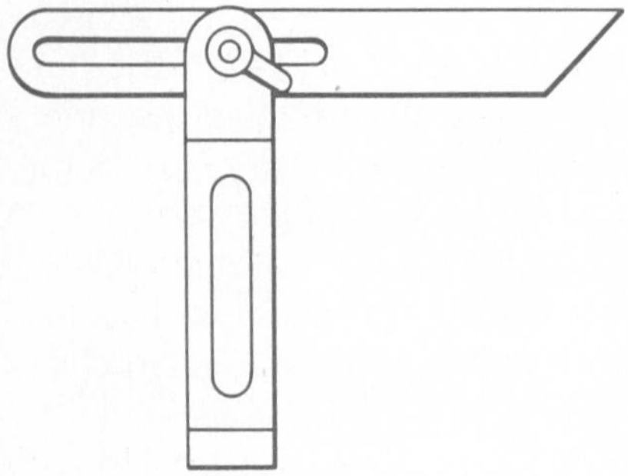

Protractor

The *protractor* is a semicircular tool usually made of either metal or plastic. It can be used to measure or mark any angle from 0° to 180°. The larger the protractor, the larger the measuring scale. A larger measuring scale ensures more accurate measurements.

A protractor also can be used to help make an angle cut when a bevel square is not available. If a cut is to be made at a 45° angle, then the protractor is placed with

Figure 8-8. A sliding bevel square is used to transfer angles.

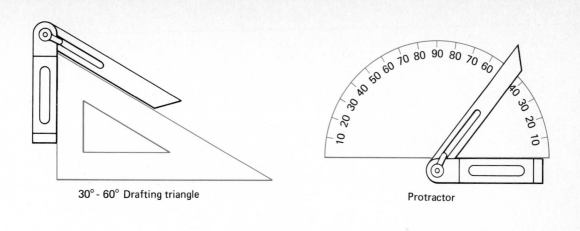

30° - 60° Drafting triangle

Protractor

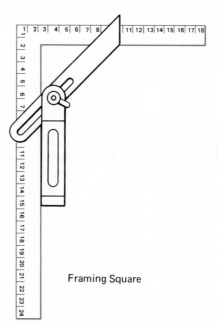

Framing Square

Figure 8-9.   Three methods used to set the bevel.

the base along the edge of the board. The arrow is then located at the point where the angle cut is to start. A mark is made at the arrow and also on the board at the 45° point on the protractor scale. With a straight-edge a line is drawn to connect both marks.

Dividers

Two kinds of dividers are pictured in Figure 8-10. They generally come in three sizes: 6, 8, and 12 in [15.2, 20.3, and 30.5 cm]. Note the thumbscrew on one leg of the divider and the nut located where the arc crosses the other leg. The thumbscrew is used for making initial adjustments of the legs. The nut is used to make the final adjustment for the most accurate mea-surement.

The dividers are used to help duplicate measurements on different pieces of wood. For example, it may be desirable to dupli-

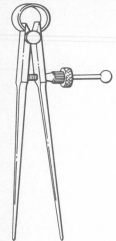

6-in flat leg divider

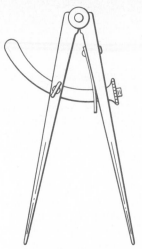
7-in winged divider

Figure 8-10. Two kinds of dividers used to duplicate measurements and mark circles or arcs.

cate a notch in other pieces of wood. By setting the legs of the dividers, it is possible to get the exact dimensions of the original notch. The dividers then can be used to mark other pieces of wood.

However, most often the dividers are used to mark circles or parts of circles (arcs).

Using Handsaws

Handsaws are the most common cutting tools for wood used by agricultural workers, whether those workers are on a farm or ranch or in a shop or food-processing plant. The handsaw is used to cut larger pieces of lumber into smaller pieces, according to the needs of specific jobs.

Crosscut Saw

A *crosscut saw* is used to cut wood across the grain. Saws generally are categorized according to the number of points or teeth they have per inch of blade. A 10-point saw, for example, has 10 teeth for every

inch of blade. The more points per inch, the finer or smoother the cut made by the saw. A 10-point crosscut saw is the one most commonly used in agricultural industry. However, they also are available in 9 and 12 points.

The crosscut saw is held by the handle, with the index finger extended along the side of the handle to help guide the saw (see Figure 8-11). The saw is held at a 45°

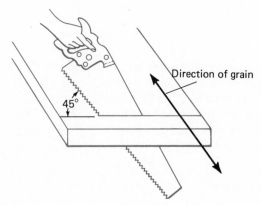
Direction of grain

45°

Figure 8-11. Correct method of holding and positioning the crosscut saw.

WOODWORKING AND BASIC CARPENTRY

angle to the wood being cut, and the other hand is used to help steady the material.

A cut made by a saw is called a *kerf*. To start the kerf with the saw, it is important for the blade to be on the side of the line that will become the scrap piece of lumber. Then the saw is drawn backward in a smooth stroke. Long, smooth forward and backward strokes are then made, following the line drawn. The blade must always be on the side of the waste lumber.

The last few strokes are made very slowly and with slightly less pressure. This helps prevent the board from splitting as the cut is completed.

Rip Saw

Ripping a board means to cut it lengthwise along the grain to make strips of narrower width. The rip saw (see Figure 8-12) is used for this work. Its teeth are coarser, with 5½

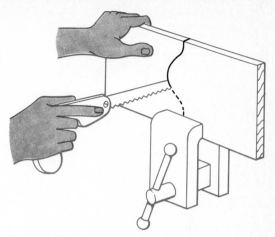

Figure 8-13.   A compass saw is used to cut curves, irregular shapes, and circles.

points per inch being the most common. The teeth actually chisel the wood rather than cut it as the crosscut saw does. Because of this difference, the saws are sharpened differently.

The rip saw works best if held at an angle of about 60° to the material being cut.

Compass and Keyhole Saws

Compass and *keyhole saws* (see Figure 8-13) are used basically for the same purposes: cutting curves, circles, and irregular shapes in pieces of lumber. The keyhole saw should be used when space is limited or the cutting radius is very short. Before a circle or other shape can be cut in wood with one of these saws, it is necessary to drill a hole large enough for the end of the blade to penetrate and start the kerf.

Coping Saw

The *coping saw* also is used for cutting curves or making other irregular cuts in wood. But its shape is much different from those of the compass or keyhole saws (see Figure 8-14). To begin with, the blade is

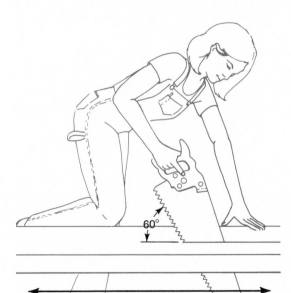

Figure 8-12.   Correct method of holding and positioning the rip saw.

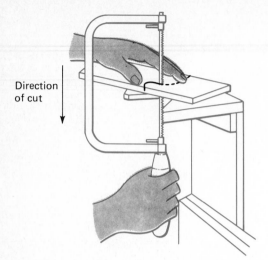

Direction of cut

Figure 8-14. Finer cuts or irregular shapes are made with a coping saw.

thin and removable. There are a great many more teeth along the blade, so the coping saw can make a much finer cut. Cuts with the coping saw should be made on down strokes only, as shown in Figure 8-16. This puts pressure on the handle rather than on the frame.

Using Hand Planes

Four kinds of planes may be found in most agricultural shops. They are the *jointer*, *jack*, *smoothing*, and *block planes*. The jointer plane is the largest, and the block plane is the smallest (see Figure 8-15). The jointer plane is usually 22 in [55.9 cm] long, the jack plane is 11 to 14 in [27.9 to 35.6 cm], the smoothing plane is 8 to 9 in [20.3 to 22.9 cm], and the block plane is 6 to 7 in [15.2 to 17.8 cm] long.

The longer the body of the plane, the more even or flat the surface being planed will become. The reason is that the larger plane tends to overcome unevenness in the wood.

The procedure used when assembling

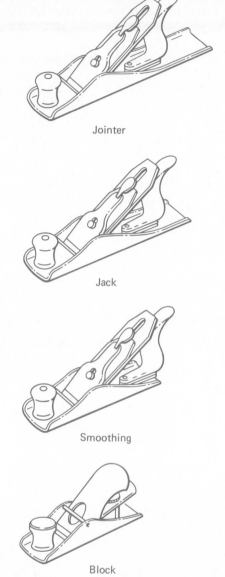

Jointer

Jack

Smoothing

Block

Figure 8-15. Four types of hand planes.

the plane (as shown in Figures 8-16 and 8-17) is very important. The plane that is improperly assembled will do inferior work. The *plane iron*—the sharp blade that does the cutting—is adjusted until the cutting edge just projects through the bottom

WOODWORKING AND BASIC CARPENTRY

Turn plane iron cap to the right.
It passes back of cutting edge

Slide plane iron cap to within
1/32 in (.79 mm) of blade edge

Bevel

Cap iron screw

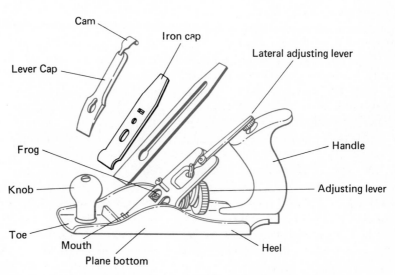

Cam

Iron cap

Lateral adjusting lever

Lever Cap

Handle

Frog

Adjusting lever

Knob

Toe

Mouth

Heel

Plane bottom

Figure 8-16.   Procedure used to assemble and adjust the plane.

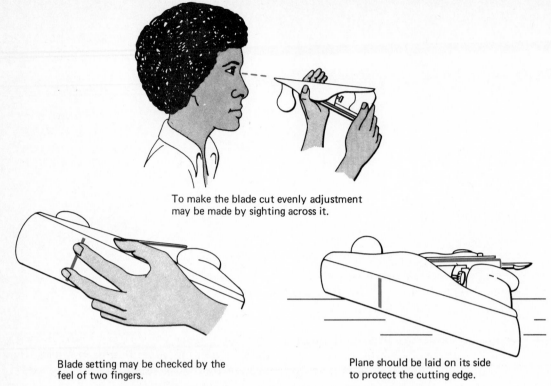

To make the blade cut evenly adjustment may be made by sighting across it.

Blade setting may be checked by the feel of two fingers.

Plane should be laid on its side to protect the cutting edge.

Figure 8-17. Procedure used to adjust the depth of the cut.

of the plane (also called the *shoe*). A lever for adjustment is used to square the plane iron with the shoe.

The farther the cutting edge projects through the bottom, the more material the plane iron is able to reach. The amount of cutting edge exposed should be adjusted according to the hardness of the wood being planed. A wood such as oak, for example, may require less blade projecting through the shoe than a softwood such as pine.

The plane is used by applying pressure to the knob and handle and moving the plane firmly and smoothly in a forward direction only (see Figure 8-18).

When not in use, the plane should be rested on its side to prevent damage to the cutting edge. If redressing (sharpening and reconditioning) is necessary, the proce-

dures outlined in Chapter 23 should be followed.

Using Wood Chisels

Chisels are designed to make special kinds of cuts in wood, such as grooves and notches. They also are used to shape and trim wood.

The chisel consists of a blade and handle. The cutting edge of the blade is beveled, or angled, in much the same way as the blade of a plane. The opposite end of the blade is called the *tang*. This end fits into the handle (see Figure 8-19). A *tang chisel* is one where the tang extends into a wooden or plastic handle. In a *socket chisel* the handle is fitted into a socket formed by the metal in the chisel.

WOODWORKING AND BASIC CARPENTRY

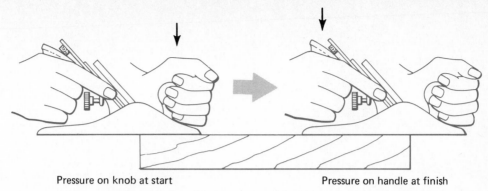

Pressure on knob at start   Pressure on handle at finish

Figure 8-18. Points where pressure should be applied when using the hand plane.

The width of chisel blades varies in size from ⅛ to 2 in [3.2 to 50.8 mm]. Chisel lengths also vary. Long-blade chisels are called *firmer chisels*, medium lengths are *pocket chisels*, and short lengths are *butt chisels*.

Chisels are used by workers by pushing the chisel forward, with the bevel (angle) down and the chisel positioned at an angle to the wood. The chisel also can be driven forward by tapping the handle or socket with a mallet or hammer. The degree of the angle usually determines how much wood is to be chiseled in one forward motion. If the chisel is held at a high angle, more wood will be cut away. There also is

danger of gouging when the chisel is held at a high angle.

Making Grooves and Notches

Common uses for a chisel are making dadoes, gaining, and rabbeting. A *dado* is a groove that is chiseled across a board. It may be desirable, for example, to chisel dadoes in the construction of shelving for a shop or equipment dealership. The grooves (dadoes) chiseled in the side panels enable the shelves to slide in and hold firmly.

Gaining is chiseling a notch in one piece of wood so that another piece can be fastened securely to it. For example, the nursery worker may want to hinge a door on a greenhouse. After drawing the outline of the hinge on the door edge, the worker would chisel out just enough wood to allow the hinge to be fastened flush to the edge of the door.

A *rabbet* also is a groove, but it is made usually along the edge or end of a piece of wood. An example of a rabbet is the groove made in a window frame to hold a pane of glass. Rabbeting is used in the construction of cabinets and in some paneling. If one panel has a lip and one panel has a groove (rabbet), when the two panels are

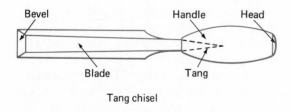

Bevel   Handle   Head

Blade   Tang

Tang chisel

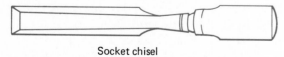

Socket chisel

Figure 8-19. Types and parts of wood chisels.

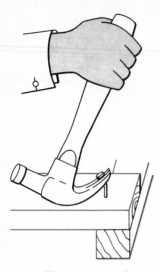

First — raise handle
to vertical position.

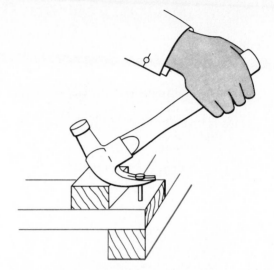

Finish pulling nail with block.

Figure 8-20.   Correct method of pulling out nails using the curved-claw hammer.

joined, the lip fits into the groove to ensure a snug fit.

Using Nail Hammers

There are two kinds of nailing hammers: one has a curved claw, and the other has a straight claw. They are often referred to as a *curved-claw* and a *ripping-claw hammer*. The curved claw is particularly valuable for pulling out nails (see Figure 8-20). The ripping claw is best used to rip apart or tear down wooden structures.

Hammers also have two kinds of striking faces. One is flat, and the other is slightly rounded. The slightly rounded head, called a *bell-faced hammer*, is used primarily to drive a nail slightly below the surface of wood without leaving hammer marks.

Nail hammerheads generally range in weight from 13 to 20 ounces [0.37 to 0.6 kg]. The 16-oz [0.5-kg] hammer is the most popular in the school shop.

When the hammer is used, it is held firmly in the hand to drive the nail. The handle is held near the end. It is important not to choke up on the handle (see Figure 8-21). Nail hammers are designed to drive unhardened nails. They are not to be used for striking hard objects such as concrete, masonry nails, or other steel tools.

When it is important to have a finished look, the head of the nail should be struck slightly below the surface of the wood. This is done by using a nail set, the only other metal object that should be struck by a nail hammer (see Figure 8-22). The nail head is driven about $1/16$ in [1.6 mm] below the surface of the wood. The little hole left may be filled with wood putty. When the putty is hardened, the surface may be sanded and finished.

Using the Brace and Bit

There are a great many occasions when holes must be drilled through wood. For

WOODWORKING AND BASIC CARPENTRY

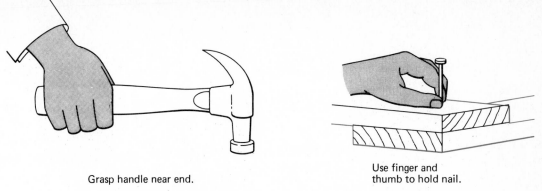

Grasp handle near end.

Use finger and
thumb to hold nail.

Figure 8-21.   Correct technique for driving a common nail.

example, a hole must be made so that the compass or keyhole saw can be started. It is common, of course, to drill *pilot holes* for screws. Another use for the brace and bit is to make holes that will fit different-sized doweling (round pieces of wood).

The main parts of the brace (see Figure 8-23) are the head, handle, and chuck. The *chuck* consists of jaws that are regulated by turning the chuck with the hand. When the jaws are open, the tang of the auger bit is inserted. The jaws are then closed to lock the bit firmly in place.

The handle must be turned clockwise to drill the hole. It is important when using the brace and bit to hold it perpendicular to the wood (straight up and down) unless the plans specify that the hole should be made at an angle.

The size of a brace is determined by the diameter of the sweep, or circle, made by the handle as it is turned. A 10-in [25.4-

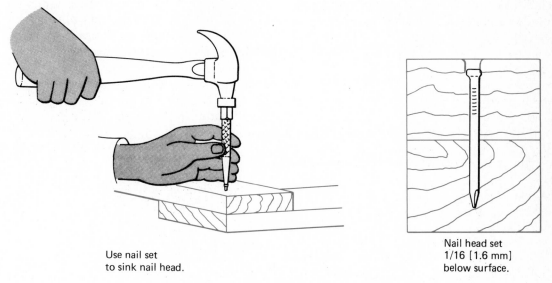

Use nail set
to sink nail head.

Nail head set
1/16 [1.6 mm]
below surface.

Figure 8-22.   Using a nail set for a finished look.

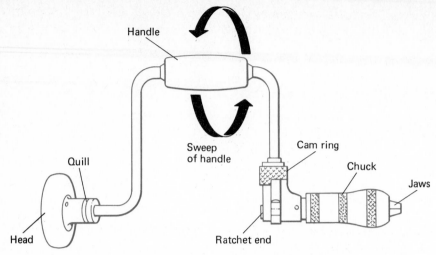

Figure 8-23.   Parts of the hand brace.

cm] brace is the one most commonly used. There are two kinds of auger bits. One has a solid shaft; the other is a flat piece of metal that has been twisted to form the flutes (grooves) of the bit (see Figure 8-24).

Generally, there are 13 auger bits in a set, starting with $^4/_{16}$ in [6.4 mm] and increasing by sixteenths to 1 in [25.4 mm]. The size of the bit is stamped on the tang. A number 4 on the tang means the bit is sized $^4/_{16}$ (or $^1/_4$ in); a number 10 bores a hole $^{10}/_{16}$ (or $^5/_8$) in [15.9 mm] in diameter.

Other Kinds of Bits

Other kinds of bits used with a brace are the expansive bit, the screwdriver bit, and the countersink (see Figure 8-25). The *expansive bit* is used to drill holes larger than 1 in [2.5 cm]. The *screwdriver bit* is used to drive (turn) screws or bolts that are slotted and are normally driven by a screwdriver. The screwdriver bit used should have a blade the same width as the slot in the head of the screw or bolt.

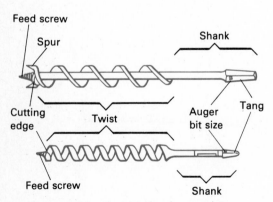

Figure 8-24.   Two kinds of auger bits.

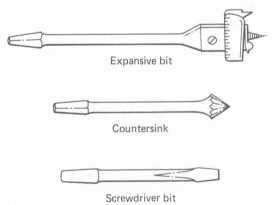

Figure 8-25.   Other bits used with the hand brace.

WOODWORKING AND BASIC CARPENTRY

The *countersink bit* is used after a hole has been drilled for a flathead screw but before the screw is driven in. The countersink bit makes a "seat" for the screw so that the flathead will be flush with the wood after the screw is driven in. One must be careful, however, not to make the seat too deep. Otherwise, the screw head will be below the surface of the wood instead of flush with it.

Bits Can Be Used on All Woods

The brace and bit can be used on any kind of wood, no matter how hard. It is possible to drill without applying a great deal of pressure, providing the bit has been properly fitted and sharpened. There are bits with a coarse-threaded feed screw for drilling in softwoods and a fine-threaded feed screw for hardwoods (see Figure 8-26). The feed screw should be allowed to advance the bit through the wood as the brace is turned. Only minimum pressure should be applied to the head of the brace when starting the hole. When the feed screw comes through the wood, it no longer advances the spur and cutting lips. The bit should not splinter the wood around the hole, therefore, because it will stop cutting. The bit can then be started from the opposite side to complete the drilling of the hole.

Coarse feed screw

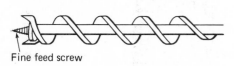

Fine feed screw

Figure 8-26. A bit with a coarse-threaded feed screw is used in softwoods and one with a fine-threaded feed screw is used for hardwoods.

Using Other Drills

The hand drill and electric drill are other tools used for boring holes in wood (see Figure 8-27). Bits for the electric drill operate at high speeds and are different from bits used in the hand drill and the brace. A discussion of the electric drill may be found in Chapter 9, Using Power Woodworking Tools.

The electric hand drill generally is not used when a high-quality finish such as a workbench top is desired. The electric drill, using flat wood bits, does not drill a very neat hole.

Using Screwdrivers

Screwdrivers come in all sizes and shapes and are designed to fit the heads of a variety of screws. Whenever materials are to be joined together with screws—and that happens in nearly all agricultural career fields—the screwdriver is needed to drive the screws.

There are basically two kinds of screwdrivers for woodworking—the standard and the Phillips (see Figure 8-28). The *standard* head is flat and tapered. The *Phillips screwdriver* has a cross-shaped bit. The screwdriver used depends on the head of the screws to be driven. Refer to Chapter 7 to review kinds of screws and how materials should be prepared to receive screws. Also, Chapter 24 discusses the care of the screwdriver. No tool is perhaps abused more than the screwdriver. It should not be used other than to drive screws.

Hand Tool Safety

When not in use, all tools should be stored in cabinets or tool boxes. This prevents them from being damaged, from becoming

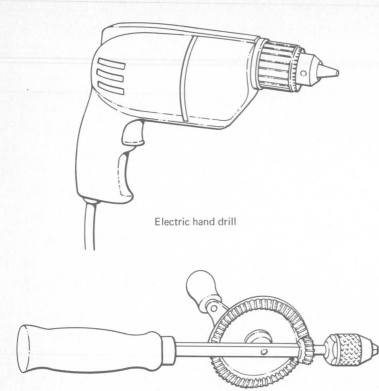

Electric hand drill

Hand drill

Figure 8-27. The electric drill and hand drill are other tools used for boring holes in wood.

dirty, and from being used improperly by unauthorized persons. Tools with handles should have the handles firmly secured and in good condition. A cracked handle on a chisel, for example, could break while being used, and the pointed tang could inflict serious injury.

Tools that use blades or other kinds of cutting edges must be kept sharp for best

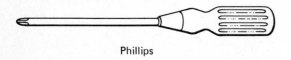

Phillips

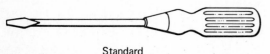

Standard

Figure 8-28. Standard and Phillips screwdrivers.

performance. The proper tool always should be selected for the job and used for only the intended purpose. A screwdriver, for example, should not be used as a chisel or pry bar.

Screwdrivers, bits, chisels, and other pointed or sharp tools should be kept out of pockets. Not only can they damage clothing, but they can also cause injury. After working, the area should be cleaned. Wood that can be used again should be put away in a proper storage rack. Waste material should be swept up and disposed of in a container set aside for that purpose.

Safety Rules

These safety rules should be followed when you use hand tools:

1. Use only tools that are sharp and in proper working order.
2. Screwdrivers are not to be used as chisels, pry bars, punches, or paint paddles. They should be used only to drive screws.
3. All handles on tools should be securely fastened before they are used.
4. Tools should be returned to proper storage and not be left lying around.
5. Select only the proper-size tool for the job. Wrong-size tools can be dangerous.
6. Keep tools out of pockets as they can create a safety problem.
7. Clean tools correctly after using. Greasy or dirty tools are dangerous to use.
8. Tools with tangs should never be used without the correct handle.
9. Keep the work areas and floor free of scrap materials.
10. Never throw tools to someone asking for a tool.
11. All minor injuries should be properly treated and reported.
12. Always wear proper eye protection when you are doing any job in the shop.
13. Horseplay is not to be permitted in the shop area.

Using Hand Woodworking Tools: A Review

Knowing how to use hand woodworking tools makes it possible for agricultural workers to take the materials they have selected and put them together according to plans sketched or drawn. These tools are used by workers in the mechanic's shop, the landscaping and nursery business, on the farm and ranch, in the food-processing plant, and in the service and supply store.

Almost all workers will need to use hand tools at some time to measure, mark and cut wood, hammer nails, drill holes, and drive screws.

It is important to remember that each tool has a specific use. Therefore, the tool must be selected carefully. Because any tool can be dangerous when it is not properly cared for or not properly used, every worker needs to follow strictly the rules for safe use.

THINKING IT THROUGH

1. Describe a situation where it is important for an agricultural worker to use measuring tools. List the problems that might arise when the measurements are incorrect.
2. How can you tell a crosscut saw from a rip saw?
3. What safety practices should be followed when you are using a hand plane?
4. What safety practices should be followed in using a nail hammer?
5. Identify and describe the use of the common hand tools in your school or home shop.

CHAPTER 9

USING POWER WOODWORKING TOOLS

Today's grandparents could describe what it was like when there were few or no power woodworking tools. All cutting, for example, had to be done with a handsaw or an ax, with nothing but muscle power—and plenty of it.

Not only did simple hand tools make the work harder, but the work also took much longer. At the modern lumber yard, workers use large circular power saws to cut lumber to the customer's specification. It is difficult to determine how long that same process took in the days when there were no power saws.

Power saws, drills, jointers, planes, and sanders have made it possible for more work to be done by fewer persons with much less time and effort.

It's logical, of course, that tools would be powered in this day when mechanization has made such an impact on agricultural industry in every other way. Agricultural workers, despite their career field, need to complete construction tasks dictated by plans as quickly and efficiently as possible. Power tools help make that possible.

The tools described in this chapter are generally found in the school vocational agriculture shop, but they also are found at many job sites in agricultural industry. The farm and ranch worker uses some or many of the tools, and the worker who specializes in construction of agricultural buildings uses most of or all the tools. The person who works in a store selling agricultural tools and equipment usually has to be familiar with these tools and prepared to demonstrate them to customers. During the off season, workers at nurseries and landscaping businesses may use power woodworking tools to construct or repair a greenhouse, equipment shed, or other building. The mechanic working in the food-processing business may use some of these tools to make or repair wood products used in the plant.

CHAPTER GOALS

In this chapter your goals are:

- To demonstrate for your instructor the safe operation of all the power woodworking tools, such as the tilting arbor saw, jointer, power handsaw, radial arm saw, planer, sander, and drill
- To pass a safety test prior to using any of the power tools

WOODWORKING AND BASIC CARPENTRY

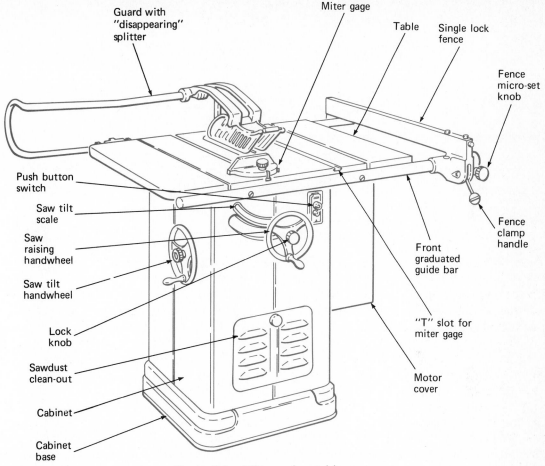

Figure 9-1. Tilting arbor table saw.

- To demonstrate what safety precautions you should observe while using each of the power woodworking tools
- To recognize when power woodworking tools are in safe and proper working condition

Using Circular Saws

There are three basic power woodworking saws. They are all called circular saws, but each has a special use. These circular saws are the *tilting arbor saw*, the *radial arm saw*, and the *portable electric saw*. The saws are illustrated in Figures 9-1, 9-2, and 9-3.

The power circular saw is a very valuable tool to have in the farm and ranch shop, in the food-processing plant shop, and at other work sites in agricultural industry. It allows the worker to cut materials fast and accurately. But the power saw also is one of the most potentially dangerous tools in the shop. Careless use of the saw can cause severe injury to the user and possibly to other persons in the shop. Safety tips for operation of circular saws are given at the end of this section.

The basic uses of power saws are to

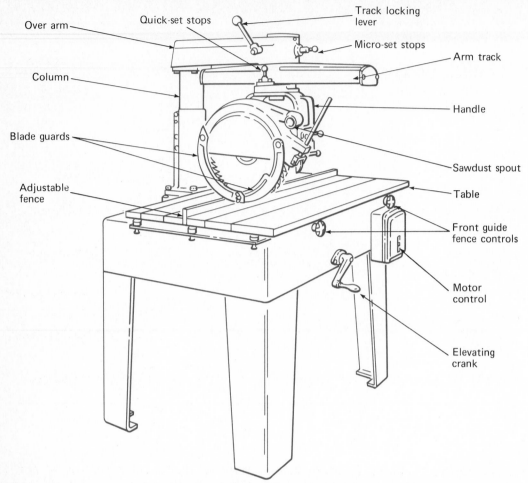

Figure 9-2. Radial arm saw.

cut lumber to desired lengths and widths and to make special cuts such as the rabbet, dado, and bevel.

The size of a power circular saw is basically determined by the diameter of the blade used. A table saw equipped with a 12-in [30-cm] blade, for example, is called a 12-in table saw.

Kinds of Circular Saw Blades

The kinds of circular saw blades used most often in the school shop, on the farm and ranch, and in agribusiness are the crosscut, rip, and combination blades (see Figure 9-4).

As the name implies, the *crosscut blade* is used to cut across the grain of wood. It is not used, therefore, to make cuts with the grain. That is *ripping*.

Cutting across the grain, of course, generally puts more stress on the wood. There is more danger of splintering. Therefore the teeth of the crosscut blade are designed differently from those on the ripping blade. The teeth of the crosscut

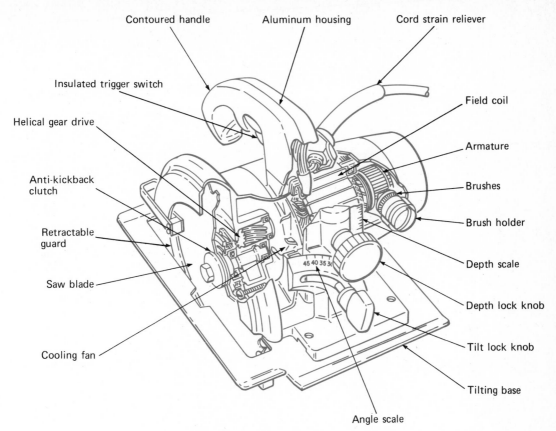

Figure 9-3. Portable electric saw.

blade are pointed and very sharp. The teeth of the ripping blade, on the other hand, do not come to a sharp point (see Figure 9-5). The rip saw blade has squared teeth that have the effect of chiseling the wood.

The name *combination blade* also implies its use. The blade consists of both crosscut and ripping teeth (see Figure 9-4). The blade can be used for both crosscutting and ripping, but the combination blade will not do either job as efficiently as the crosscut or ripping blades separately.

The crosscut, ripping, and combination blades are those most often used, but not all crosscut, ripping, and combination blades are alike. For example, if a worker

were going to cut plywood or particle board for sheathing the inside of a storage building or paneling the salesroom of a nursery, a blade (or blades) would be used with more and smaller teeth than if the worker were cutting rough-sawn timber for an agricultural storage building.

Before any job is undertaken, workers must make sure they have the best blade to do the job. When they are uncertain about which blade to use, workers should check the manufacturer's recommendations and talk to the person selling the building materials. This is why it is so important that sales clerks and others working in agricultural supplies/services know the power tools they sell and service.

Type of blade		Diameter	Arbor hole
Combination saw blades (flat ground)		12″	1″
		14″	1″
Rip saw blades (flat ground)		12″	1″
		14″	1″
Crosscut blades (flat ground)	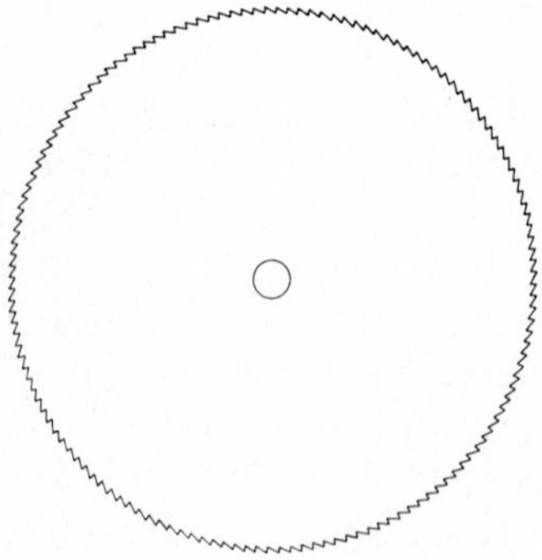	12″	1″

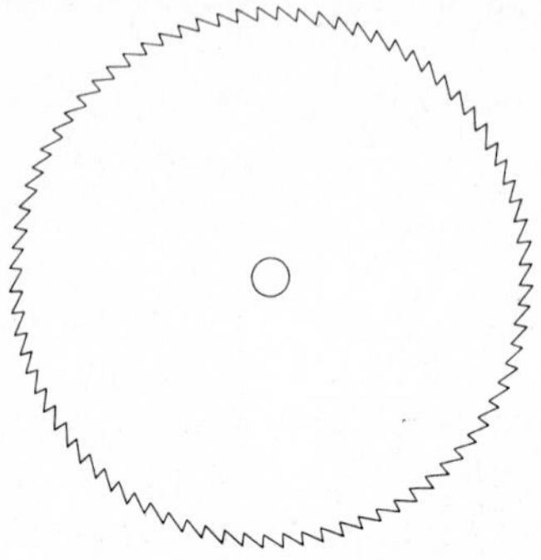

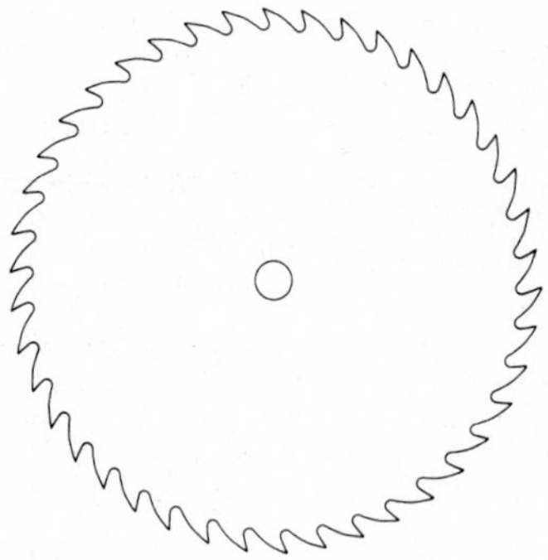

Figure 9-4.   Blades commonly used on the circular saw.

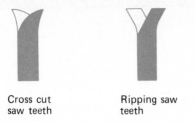

Cross cut
saw teeth

Ripping saw
teeth

Figure 9-5. Shape of the cross cut and ripping saw teeth.

Using the Tilting Arbor Saw

Figure 9-1 shows the various parts of the tilting arbor saw. It is particularly important to know how to adjust two parts of the saw: the *miter gauge* and the *ripping fence*.

Miter Gauge. The miter gauge holds wood for crosscutting at any angle between 45° and 90°. The miter gauge is adjustable and can be set to a desired angle by using a square or a bevel square (see Figure 9-6). The miter gauge fits either the right or the left groove ("T" slot) in the tabletop. For most operations, the left groove is more desirable for the worker who is right handed. The material to be sawed is held firmly against the gauge and pushed straight forward into the saw blade.

Ripping Fence. The ripping fence is used to cut material into different widths. It is never used for crosscutting. One reason it is not used for crosscutting is that a small piece of waste wood could get stuck between the fence and the saw blade and then be kicked back with enough force to seriously damage the saw and injure the operator or someone else in the shop. The fence can be checked for squareness with the tabletop by measuring the distance from the fence to one or both miter gauge grooves.

When very narrow or short pieces are being ripped, a push stick should always be used to protect fingers from the saw blade (see Figure 9-7). In ripping a long board, it helps to have a partner to hold the other end as it emerges from the saw blade and begins to extend over the table. If there is no one available to help, a support called an off-feed roller that is the same height as

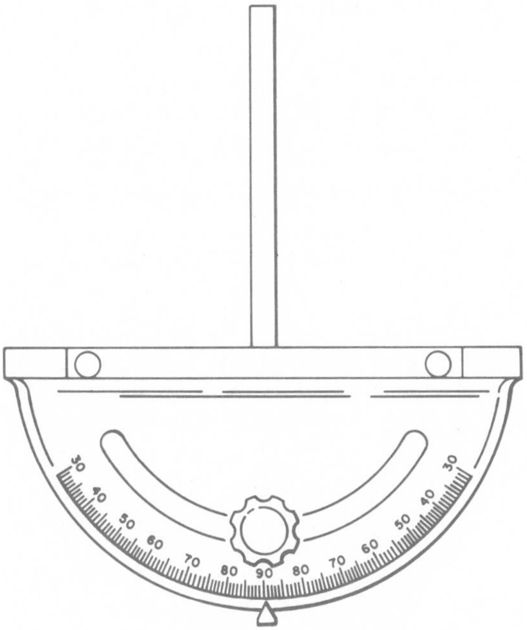

Figure 9-6. Miter gauge for the tilting arbor table saw.

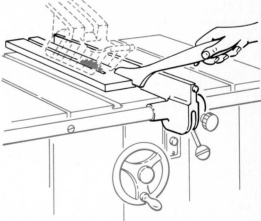

Figure 9-7. A push stick should be used when ripping narrow or short pieces of wood.

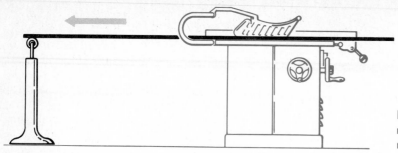

the table can be used (see Figure 9-8). The operator should never attempt to reach over the moving blade to guide the lumber as it comes from the saw.

The saw can be used for special cuts such as the dado, rabbet, and bevel.

Dado Cut. The dado is a wide groove that may go across the width of a board. The dado is particularly important when strength of the joint is important. The dado is cut into the upright, and the cross-member fits into the groove. There is a special dado saw head (see Figure 9-9). The head consists of two outer blades and several spacing or chipper blades. The two outside blades usually are ⅛ in [3.2 mm] thick, and the chipper blades vary in width from 1/16 to ¼ in [1.6 to 6.4 mm]. The width of the dado cut can be further adjusted by placing thin washers or spacers made from various thicknesses of paper between the chipper blades.

Rabbet Cut. A rabbet also is a groove, but this groove is cut along the outer edge of a board or down its length. The rabbet is used, for example, to connect boards when you are paneling. The lip on the edge of one board fits snugly into the rabbet along the edge of the next board. Material called *shiplap* or *car siding* is often used to construct feed bins. The rabbet cut is made with a circular saw blade, by using the ripping fence as a guide. The saw blade, of course, is set at a shallow depth (determined by the desired depth of the groove). The instructor can demonstrate how a rabbet groove is cut.

Bevel Cut. The bevel cut is made at an angle. It may be desirable, for example, to cut the edge of a board at a 45° angle instead of square. The desired bevel is determined by using the miter gauge. The miter gauge will position the board at the desired angle to the cutting blade. The in-

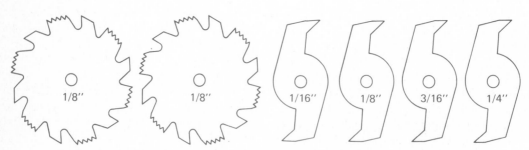

Figure 9-9. A dado head set includes blades and chippers of various widths that form the dado joint.

WOODWORKING AND BASIC CARPENTRY

structor can demonstrate how a bevel cut is made.

It is important to have the instructor make any adjustments or repairs to the saw.

Using the Radial Arm Saw

The radial arm saw (see Figure 9-2) has some advantages over the tilting arbor saw. The blade is above the table instead of coming up from below the table. This allows the operator to see more clearly how the work is progressing. Also, long boards can be crosscut by the radial arm saw without the problem of waste wood being kicked back. The blade is attached directly to the electric motor shaft, thus eliminating the need for belts and pulleys.

When the radial arm saw is used to crosscut, the blade is adjusted so that it just touches the top of the table. While this adjustment or any other blade adjustments are being made, of course, the electric cord should be disconnected and the arm with the blade should be in the back, or rest, position against the column.

The board is placed against the fence and held firmly with one hand while the other hand pulls the blade slowly through the board. When making this cut, the operator should stand slightly to one side of the blade. After the cut has been made, the blade is returned to the column. If no other cuts are to be made, the motor should be turned off.

Bevel, or angle, cuts can be made by adjusting the angle of the saw blade. The blade can be angled to either the left or the right, largely depending on the position and location of the angle that is required. Generally, the machine is angled to the right because it has greater capacity.

Students using the saw should carefully follow the teacher's instructions and the directions given in the operator's manual.

When the radial saw is used for ripping, it is vital that the lumber to be cut be started from the correct side of the saw. First adjust the saw blade so that it is parallel to the fence at the back of the table. The blade can be turned either way (for an "in-rip" or "out-rip"), depending on the thickness of the finished piece required. An out-rip is preferred because the blade is more visible to the operator. The material is fed into the saw from the side where the blade teeth are on the up-cut and *not on the down-cut*. The saw should have a decal or sign on the guard which says, "Rip from this side only." This is a most important safety feature; the consequences of not following it can be very serious. If material is fed from the wrong direction, the teeth may cause the blade to act as a wheel and throw the material out of the saw.

The saw is equipped with anti-kickback fingers that should be used at all times in ripping. After the rear of the saw blade guard is adjusted so that it almost touches the board, the antikickback fingers should be adjusted so that they dig into the material if kickback occurs. This will prevent a piece of wood from kicking back into the operator or someone else (see Figure 9-10).

Using the Portable Electric Saw

The portable electric saw is one of the most commonly used tools in agricultural industry. In fact, it is a common tool in most other industries and in most homes in the United States. It has many uses other than cutting lumber. When equipped with the proper blade, for example, it can be used to cut metal, masonry products, glass, ceramics, and even some concrete products. The parts of the portable electric saw, as shown in Figure 9-3, should be studied.

Most portable electric saws are powered by a universal electric motor with a saw blade attached to the motor shaft. If

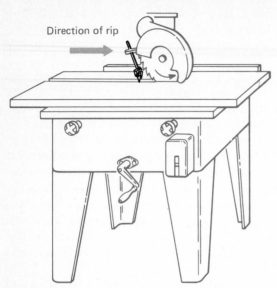

Direction of rip

Figure 9-10. Adjust the blade guard and anti-kickback when ripping.

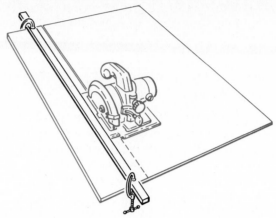

Figure 9-11. Using a guide strip to cut large pieces.

the sawing is to be done on lumber that is lying flat, the lumber should be properly supported so that waste wood will drop off instead of falling against the blade.

The base of the saw is placed on the lumber, with the blade aligned with the mark made for the cut. Before the power is turned on, the operator should make sure that the blade is not touching the lumber. If it is, the blade may gouge the wood and stall the motor.

As with the other saws, the size of the portable electric saw is determined by the diameter of the blade.

The base of the saw is adjustable for both depth and angle (bevel). The blade and operator are protected by an upper and lower guard. The lower guard is retractable and pushes back away from the lumber as the cut is being made. This guard should be checked to make sure that it is free to swing up and down.

If a straight piece of lumber or angle iron can be clamped along the mark on the

lumber to be cut, this will help guide the saw (see Figure 9-11).

Be sure the blade has completely stopped revolving before you set down the saw anywhere. The instructor should make any repairs on or adjustments to the saw.

Safety Rules

These safety rules should be followed when circular saws are used:

1. Notify the instructor if you intend to practice on any power saw.
2. Turn off the main power by disconnecting the power cord to the saw before changing blades or making major adjustments. This prevents the saw from starting accidentally if a switch is bumped.
3. Make sure all guards on the saws are in their proper positions and working correctly.
4. Always stand to one side when you are using a saw; never stand directly in front of the blade.
5. Use the miter gauge and/or fence to guide the cutting.

6. When you are using the portable saw, clamp a piece of lumber as a guide.
7. Keep blades sharp; dull blades cause accidents.
8. Generally, the blade should not project through the material being cut more than ¼ in [6.4 mm].
9. Never reach over a spinning blade.
10. Always use a push stick when feeding small pieces of wood through the blade.
11. Do not try to brush away chips or sawdust from the front of a turning blade.
12. Do not leave the saw when the blade is turning.
13. Know where cutoff switches are so that you can react quickly in an emergency.
14. Keep the work area clean.
15. Loose materials and tools should be kept off the saw table.
16. Do not wear clothing with floppy or loose sleeves that may be caught up in the saw blade.
17. Watches, rings, and bracelets should be removed before you begin work.
18. Long hair should be either netted or worn under a cap to prevent it from catching in the blade.
19. Safety glasses should be worn, or a face shield should be used.
20. All lumber being sawed should be free of nails, loose knots, or other materials that may interfere with the blade.
21. Don't attempt to rip warped or crooked boards with the table saws.
22. The electrical cord should be kept out of the way when you are using the portable saw.
23. Saws should not come in contact with water, oil, or grease.
24. The portable saw should not be placed where it can be knocked to the ground or floor.

25. Do not operate any saw when you are standing in water or a damp area.
26. When ripping, feed the stock from the correct end of the table.
27. Set the blade guard to just clear the stock and adjust the antikickback fingers.

Using the Band Saw

The band saw is used in sawing curved or irregular shapes in wood. It also can make straight cuts if a ripping fence or miter gauge is used. Curved and irregular cuts generally are made freehand, without a fence or miter gauge.

A typical band saw is shown in Figure 9-12. The cutting blade is an endless steel band with teeth along one edge. The blade runs over two large wheels, or pulleys. The lower wheel is attached to the power source.

The size of the band saw is determined by the distance from the blade to the arm, or column (see Figure 9-12). The most common band saw is 14 in [35.6 cm]. This means that a cut can be made to the center of a piece of wood measuring 28 in [71.1 cm].

Band-saw blades used to cut wood are like the blades of a handsaw. The size of the teeth and the number of teeth per inch determine whether the cut will be fine or coarse. A blade with 4 to 7 teeth per inch is used in woodworking. The width of the blade determines how sharp a curve or radius can be cut. A blade ⅛ in [3.2 mm] wide will cut a smaller radius than a blade ½ in [12.7 mm] wide.

Metals also can be cut with a band saw. A metal-cutting blade has more teeth per inch, usually 10 to 24, and the blade runs at a slower speed.

Before the band saw is used, the blade should be properly installed. In going from

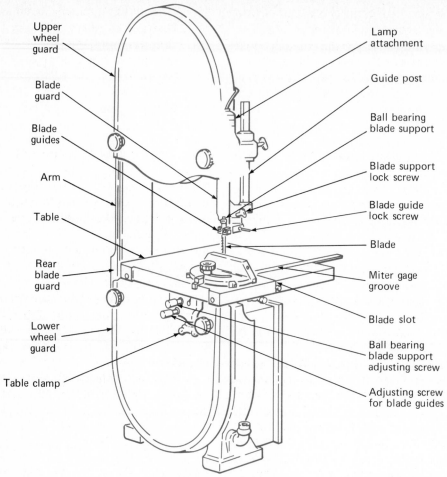

Upper wheel guard

Blade guard

Blade guides

Arm

Table

Rear blade guard

Lower wheel guard

Table clamp

Lamp attachment

Guide post

Ball bearing blade support

Blade support lock screw

Blade guide lock screw

Blade

Miter gage groove

Blade slot

Ball bearing blade support adjusting screw

Adjusting screw for blade guides

Figure 9-12. Band saw.

one blade thickness to another, the blade guides and ball-bearing support must be adjusted according to the operator's manual. Major repairs or adjustments to the band saw should be made by the instructor.

Cutting with the Band Saw

The band saw is used to cut lumber to a specific length or width and to cut an irregular shape or surface on lumber or other materials.

The tabletop should be adjusted to cut at a 90° angle or at any angle between 45° and 90°. The blade guard and hold-down mechanism should be adjusted to just clear the thickness of the material being cut. Then the band saw is turned on, with the operator standing to one side until the blade comes up to its maximum speed.

The material is cut by using the ripping fence or the miter gauge, if possible. If the material is of a size that does not permit the use of either the fence or the gauge, then the material must be cut freehand. Sharp corners in cutting should be avoided, because they will cause the blade to bind and possibly to break.

When the cut is completed, the saw should be turned off. The operator should not leave the saw until the blade comes to a full stop.

Safety Rules

These safety rules should be followed when you use the band saw:

1. When blades are being installed, the power supply to the saw should be shut off.
2. Do not brush away chips or sawdust from in front of a rotating blade.
3. Avoid backing the blade out of a cut. There should be an open exit for it. If it is absolutely necessary to back out of a cut, make sure the saw is turned off.
4. Don't use the saw unless both the upper and lower wheel guards are in place and securely fastened.
5. The blade guard located between the wheel guard and the tabletop should be adjusted to within ¼ in [6.4 mm] of the work.
6. Check blade tension and alignment by first turning the wheels by hand.
7. Use blades that are sharp and are of the proper width for the job.
8. Do not use blades that are cracked or bent or have teeth missing.
9. Keep the work area free of scraps and tools.
10. Do not wear loose clothing, rings, bracelets, or a watch. Keep long hair out of the way.
11. Wear safety glasses or use a face shield.
12. In cutting small pieces, use a stick to push the material.
13. Have the proper instructions for the job before you start it.
14. Consult the operator's manual before starting the saw.
15. Get help when cutting very large pieces of wood that overhang the table.

Using the Jointer

The *jointer* is used to remove material and to straighten and smooth the edge of a board. *It is a potentially dangerous tool!* You must be thoroughly familiar with its parts and method of operation before using it (see Figure 9-13).

The size of a jointer is determined by the length of the knife in the cutter head. The cutting knives are located on a revolving cutter head located in the center of the table. In effect, the cutting head separates the table into two parts—the front table and the rear table. Both tables are adjustable up and down. In almost all cases, the rear table is adjusted so that it is flush with the knives at their highest point (see Figure 9-14).

Removing Material with the Jointer

When the operator has determined the amount of the cut necessary, the front table is set using the depth scale. If it is desirable to trim off ⅛ in [3.2 mm], for example, the front table would be lowered until the depth scale registers ⅛ in. If it is necessary to remove more than ⅛ in, several cuts should be made.

The jointer is equipped with a fence that assists in moving the material properly across the cutting head. The fence may be adjusted to cut an angle of 45 to 90° on the edge of the board being run through the machine.

When there is a choice, the board should go through the jointer with the grain (see Figure 9-15).

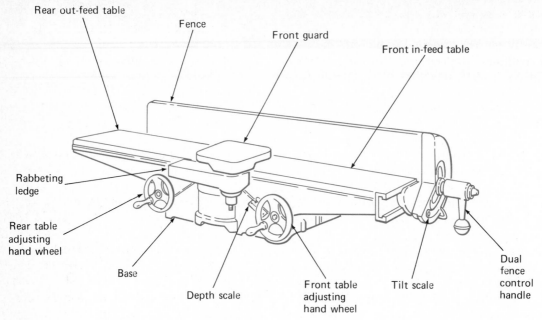

Figure 9-13. Jointer.

Straightening with the Jointer

When jointing, or straightening, the edge of a board, first the operator must make sure that the rear table is exactly level with the cutting knives. Then the front table is lowered by the amount of the cut desired. Of course, the jointer is turned off when these operations take place. The fence is checked to make sure it is properly adjusted, and the cutter guard is checked to make sure that it is working properly and correctly located.

When the power is turned on, the operator should stand to one side. Then the board is placed on the front table and held firmly against the fence. The wood is advanced carefully across the cutting head and onto the rear table. Moderate downward pressure is applied to the board while it is on the front table. As the board passes over the cutter to the rear table, downward pressure is applied. The other hand is only guiding the remainder of the board over the front table.

The operator should never allow

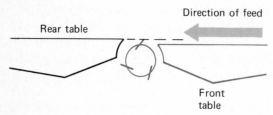

Figure 9-14. Adjusting the rear out-feed table.

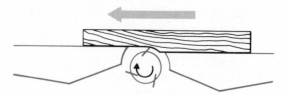

Figure 9-15. Feeding a board throught the jointer following the grain.

WOODWORKING AND BASIC CARPENTRY

hands to be directly over the cutting head. A jointer push stick is used to move material over the cutting head. The operator also should keep fingers off the front table. This is done by using the heel of the hand to push.

Boards that are less than 12 in [304.8 mm] long or 1 in [25.4 mm] wide should not be run through the jointer. The reason is that small pieces reduce the margin of safety for the operator's hands.

It is possible to joint the end of a board if the board is not too narrow. But if the width of the board is less than 6 in [15.2 cm], it should not be jointed. A narrower board may get caught by the knives and be kicked out.

When jointing the end, the board is held against the fence on the front table. The board is then pushed forward about one-half in. Next the board is turned around, and the cut is finished. By turning the piece around, the operator may avoid splintering the wood. Several shallow [less than ⅛ in (2.1 mm)] cuts are much better than one deep cut.

All major adjustments of the jointer, including knife sharpening, should be done only by the instructor.

Safety Rules

These safety rules should be followed when you are using the jointer:

1. Know the parts of the jointer and how to use them correctly.
2. Use all guards and make adjustments to the tables and fence prior to turning on the machine.
3. Make sure knives are sharp. Dull knives can cause accidents.
4. Do not joint boards that have a length of less than 12 in [30.4 cm] or widths of less than 6 in [15.2 cm].
5. Cut with the grain.

6. Keep the edge or side of the board firmly against the fence.
7. Apply downward pressure to the board as it begins to go from the front table to the rear table.
8. Hands should never pass over the cutting head. Use a push stick. Also, fingers should not drag along the table.
9. Keep the tables free of wood scraps and tools.
10. Do not run cracked boards or boards with knots or nails through the jointer.
11. Always stand to the side while wood is running through the jointer.
12. Make several shallow cuts rather than one deep cut.
13. Get assistance if long boards will be run over the machine.
14. Do not wear loose clothing, rings, bracelets, or a watch. Long hair should be kept under a net or cap.
15. Wear safety glasses.

Using the Thickness Planer

The *planer* is used to smooth rough lumber and to reduce the thickness of the wood. The size of the planer is determined by the width of the cutter knives or length of cutter head. Like the jointer, the planer has knives locked into a revolving cutter head.

Some planers have two cutting heads, but most have only one. If it is a single cutting head, it will be above the table. The table height is adjusted by turning an elevating hand wheel (see Figure 9-16).

The table is equipped with two idle rollers that help move the lumber through the machine by reducing friction between the lumber and the tabletop. There also are two power feed rollers. The in-feed roller is grooved to help force the lumber past the revolving cutter head.

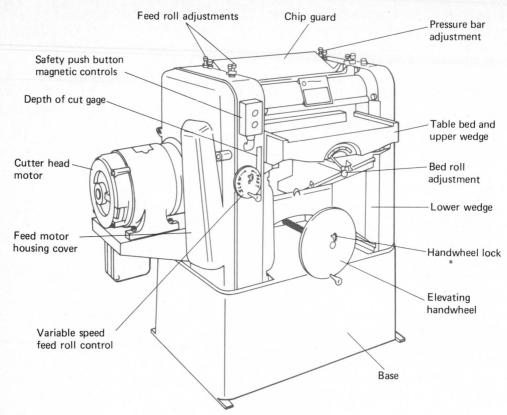

Figure 9-16. Thickness planer.

Lumber that is splintered or cracked or contains nails or large knots should not be run through the planer. As with other power equipment, the operator should stand slightly to the side when running wood through the planer.

The planer is equipped with a depth gauge that determines the amount of the cut. A cut over ⅛ in [3.2 mm] should not be attempted in a single pass. Instead, the wood should be run through the machine as many times as necessary to arrive at the desired depth. If a board is 1 in [25.4 mm] thick and the finished thickness should be ⅞ in [22.2 mm], then the depth gauge should be set to read ⅞ in.

As in the case of the jointer, it is dangerous to run pieces of wood smaller

than 12 in [30.4 cm] through the planer. Any piece that is shorter than the distance between the idle rollers should not be planed.

The planer is designed mainly to plane the flat surface of lumber. However, it also can be used to plane the edges of boards. Boards to be planed on edge should be approximately the same width and have one straight edge. If one edge isn't straight, it should be run through the jointer.

Several pieces may be clamped together and started through the planer. The clamp should be removed before it gets near the cutter head or the table. Boards with an edge width of less than 2 in [51 mm] should not be planed.

The planer is a precision tool, and it

WOODWORKING AND BASIC CARPENTRY

needs to be properly adjusted. The instructor should make all adjustments, other than raising and lowering the table.

Safety Rules

These rules should be followed when you are using the planer:

1. Do not plane boards that are less than 12 in [80 cm] long.
2. Do not make adjustments if the power is on, and never leave the machine if it is running.
3. Make sure that the wood being planed is free of nails, loose knots, and other materials that might cause problems.
4. Do not run boards of different thicknesses through the planer at the same time, because the thinner board might be kicked back.
5. Never try to remove shavings while feeding wood through the planer.
6. Keep fingers away from the in-feed table when the wood is going through.
7. Keep scraps of wood and tools away from the tabletop.
8. Get help or use special support stands when feeding through very long boards.
9. Do not wear loose clothing, rings, bracelets, or a watch when using the planer. Keep long hair in a net or under a cap.
10. Wear safety glasses.

Using the Sander

Certain kinds of woodworking jobs make it desirable for the wood to have a clean, smooth surface. Such jobs might include the construction of shelves, paneling, and the tops of workbenches in agricultural buildings and shops.

Sanders usually are of two kinds—*belt* or *disk* (see Figure 9-17). Sometimes the machine may combine the two operations. The size of the machine is determined by the diameter of the disk or the width of the belt.

The sander can be used on both wood and metal. If much metal is to be sanded, however, a special disk or belt for metal should be installed on the machine.

The revolving disk or belt moves past a stationary table that can be adjusted for sanding at various degrees. In sanding with the disk, the work should be done on the down side of the disk rather than on the up side. By sanding on the down side, the wood being sanded can be held firmly on the table.

The table is equipped with a miter gauge that permits sanding at any angle between 45° and 90°.

As with other power machines, the instructor should make any major repairs or adjustments to the sanding machine.

Safety Rules

These rules should be followed in using the sander:

1. Be sure the disk or belt is properly installed and is in good condition.
2. Do not use disks or belts that are excessively worn.
3. All guards and shields should be in their proper location.
4. Be familiar with all parts of the machine and consult the operator's manual.
5. Keep the table and area free of scraps and other tools.
6. Do not wear loose clothing, rings, bracelets, or a watch. Keep long hair under a net or cap.
7. Wear safety glasses.

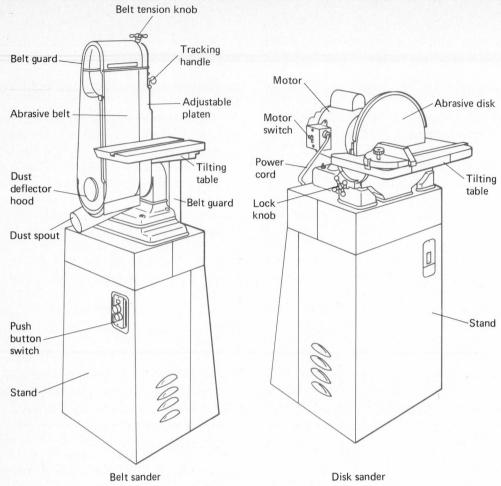

Belt tension knob

Belt guard

Tracking handle

Abrasive belt

Adjustable platen

Dust deflector hood

Tilting table

Belt guard

Dust spout

Push button switch

Stand

Belt sander

Motor

Motor switch

Abrasive disk

Power cord

Tilting table

Lock knob

Stand

Disk sander

Figure 9-17. Upright disk sander.

Using Portable Electric Tools

In addition to the portable circular saw, there are other portable tools powered by electricity. These tools are widely used by many workers on the farm and ranch and in agribusiness.

A general rule about portable electric tools is that they should be used with a properly grounded electrical system. Electrical outlets used for portable tools must be equipped with a *ground fault interrupter* (GFI). Many portable electric tools are double-insulated and are equipped with a plastic handle or body that protects the operator from shock. However, no electric tool should be used in the rain or in areas where the operator must stand in water.

Electric Drill

Electric drills come in many sizes and shapes. Their size is determined by the size of the steel twist drill the chuck will receive (see Figure 9-18). The more common sizes of drills are ¼, ⅜, and ½ in [6.4, 9.5, and

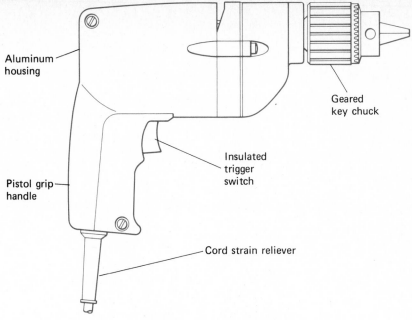

Figure 9-18. Electric hand drill.

Aluminum housing

Geared key chuck

Insulated trigger switch

Pistol grip handle

Cord strain reliever

12.7 mm]. Figure 9-19 shows a boring bit and a twist drill.

In using either the twist drill or the boring bit in wood, often it is necessary to prevent the wood from being splintered where the bit exits. One way to prevent splintering is to clamp a piece of scrap material over the spot where the bit will exit. Then the bit goes through the main piece of wood into the scrap piece. Another way to avoid splintering is to withdraw the bit just before exiting and finish the hole from the other side.

The twist drill or boring bit is inserted into the chuck of the drill. The chuck, called a *Jacob's universal chuck*, generally consists of three jaws, which are opened and closed together with a "key." To be sure that the chuck is tight and the bit or drill secure, the key should be used to tighten the three jaws. The chuck key should be removed before drilling.

Portable Sander

There are two kinds of portable sanders— the *finishing sander*, with sanding pad, and the *belt sander* (see Figures 9-20 and 9-21). The finishing sander uses a sanding pad

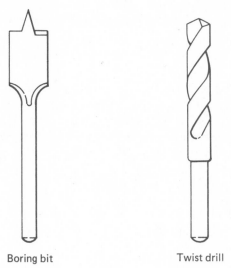

Boring bit   Twist drill

Figure 9-19. Boring bit and twist drill.

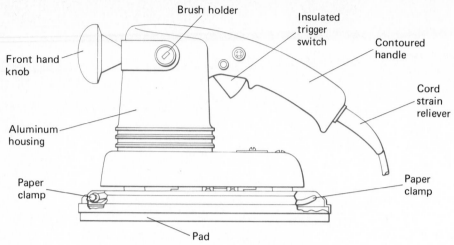

Figure 9-20. Portable belt sander.

that may either revolve in a circle (orbital) $^3/_{16}$ to $^1/_4$ in [4.8 to 6.4 mm] in diameter or go back and forth (oscillating). Most such sanders use approximately one-third of the standard 9×12 in [22.9 × 30.5 cm] sheet of sandpaper. The size of the belt sander is determined by the width of its belt. The most common width is 3 in [76.2 mm].

Before starting to sand, the worker must decide whether the job calls for a coarse sandpaper, a very fine sandpaper, or something in between. A very fine sandpaper is used for the final finishing stage. The kind of paper is determined by the mesh size of grit on the paper. A mesh size of 400 (10/0) means a very fine paper,

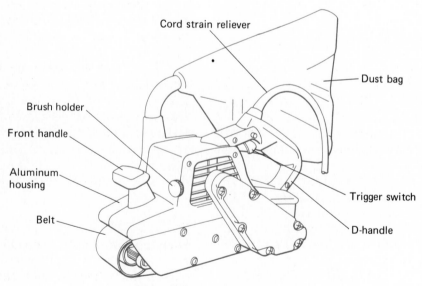

Figure 9-21. Belt sander.

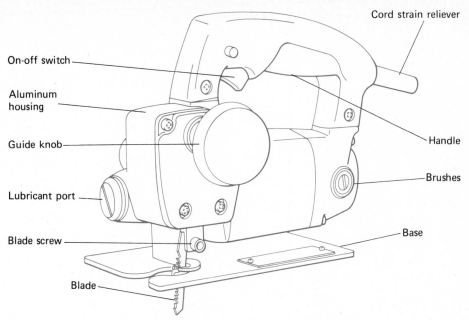

Cord strain reliever

On-off switch

Aluminum
housing

Guide knob

Lubricant port

Blade screw

Blade

Handle

Brushes

Base

Figure 9-22.   Portable saber saw.

and a mesh size of 20 (3½) is very coarse. It is generally desirable to sand with the grain.

The weight of the sanding machine is usually enough pressure on the wood. The operator merely guides the machine.

Portable Saber Saw

The saber saw, which is sometimes called the *bayonet saw*, is a very versatile tool (see Figure 9-22). By using different blades, the saw can be used to cut either straight or curved lines. Also, the saw will take blades for cutting wood, metal, and other materials.

In selecting a blade for the saber saw, a general rule is that no matter what the material is and how thin it might be, there should be at least two sawteeth in contact with the material at all times. Many saber saws are equipped with different speeds for different materials. In cutting steel, for example, a very slow speed would be used.

The saw is equipped with an adjustable base that permits bevel cuts to be made. Firm and constant pressure is applied in using the saw, but the saw should not be pushed beyond its limits. When curves are cut with the saw, a narrow blade is used to prevent binding. A wider blade can be used for straight cutting.

When it is necessary to start a cut in the middle of a piece of wood or metal, a hole should be drilled first. In very soft materials, the blade itself might start the hole, but this is not possible in most cases.

Using Power Woodworking Tools: A Review

No matter where workers are employed in agricultural industry, it is almost certain they will have to use power woodworking tools at some time to implement a construction plan, build products out of wood, or make repairs on products made from wood. The workers in food production use power woodworking tools for such jobs as construction of barns, storage buildings, and feeders. The workers in nurseries and landscaping businesses use the equipment to construct greenhouses and sheds, work-tables, and offices. The general repair shop worker may use power tools to make workbenches or repair a wooden truck floor. People employed by companies selling and servicing equipment and chemicals need to know how to use power wood-working tools so they can demonstrate them to customers. Also they may use the tools themselves in the construction of shelving and salesroom and storage areas.

There are different kinds of saws and different kinds of blades for specific jobs. It is critical to the success of the job that the right saw and blade be chosen.

Power tools can be dangerous to the operator and others if exceptional care is not taken in their use. It is extremely important, therefore, that safety rules be strictly enforced.

THINKING IT THROUGH

1. What kind of shop dress is appropriate when using power woodworking tools? Why?
2. Why do employers stress good safety habits?
3. How is the size determined on power woodworking saws?
4. What special cuts can be made using the circular or radial arm saw?
5. What special precaution should be taken when ripping with the radial arm saw?
6. What hazard exists when operating portable electric tools and how do you protect yourself as an operator?
7. What general rule determines the number of teeth per inch on the band saw?
8. What is the minimum length of material which may be used in the jointer and planer?

UNIT IV
METAL WORKING

COMPETENCIES

Competency	PRODUCTION AGRICULTURE — Contract farmer	AGRICULTURAL SUPPLIES/SERVICES — Livestock supplies services worker	AGRICULTURAL MECHANICS — Agricultural welder	AGRICULTURAL PRODUCTS, PROCESSING, AND MARKETING — Processing plant maintenance mechanic	HORTICULTURE — Groundskeeper	FORESTRY — Log skidder operator	RENEWABLE NATURAL RESOURCES — Park worker
Identify various metals used in agricultural industry	Very Important	Important	Very Important	Very Important	Important	Important	Not Important
Select and use hand tools for metal working	Very Important	Very Important	Very Important	Very Important	Important	Important	Not Important
Identify and demonstrate the correct use of power metal-working tools	Very Important	Important	Very Important	Very Important	Not Important	Important	Not Important
Demonstrate the safe use of drilling and tapping tools	Very Important	Very Important	Very Important	Important	Very Important	Not Important	Important
Demonstrate the procedure used to thread a bolt	Very Important	Very Important	Very Important	Very Important	Important	Very Important	Important
Select fasteners for cold metal	Very Important	Very Important	Important	Very Important	Important	Important	Important
Heat-treat to harden or anneal metal	Important	Important	Very Important	Important	Not Important	Important	Not Important

Legend:
- Very Important
- Important
- Not Important

114   METAL WORKING

Most workers in the various career fields of agricultural industry need to know as much about metal working as they do about woodworking.

Many things that are constructed in agricultural industry combine both wood and metal. A good example is a farm storage building that has a frame made of wood but outside siding made of metal.

Agricultural workers also need to know how to work with metal, because the machinery and equipment of agricultural industry are made primarily of metal. Metal-working skills are required in order to make the necessary repairs on machinery and equipment.

Upon completion of this unit, you should know how to identify and select various metals. This is just as important as knowing how to pick the right lumber for the job. Then you should be able to identify metal-working tools and explain how they are used. Some of the tools are the same as those used for woodworking, but with important differences such as the use of special blades or operation at different speeds. Of course, there are major tools made just for metal working that must be mastered.

Finally, you should know how to work with both cold and hot metal—how to drill, form, cut, and fasten it.

Agricultural workers with both woodworking and metal-working skills are of great value to their employers—whether they are on the farm and ranch, in the food- and nonfood-processing plant, the selling and servicing business, the mechanic's repair shop, or the nursery and landscaping business.

CHAPTER 10

IDENTIFYING AND SELECTING METALS

Today we live in the space age, but those who have studied ancient history know of the Iron Age. It was that time in human existence that made possible many of the ages since then.

The effect of the Iron Age on agricultural industry has been particularly important. Most of the farm machinery and equipment today is made of iron or a combination of iron and other materials.

Steel, for example, is iron which has been refined and combined with a desirable amount of carbon. At one time, most tools were made of wood, bones, seashells, or stones. Iron and steel, which can be shaped and heat treated, have had a tremendous impact on mechanized agricultural industry.

Iron is made from iron ore. Iron ore has been in the earth and mined for centuries. American Indians used red iron ore in combination with other natural substances to paint their faces and bodies.

Before iron ore can be used for manufacturing, it has to go through several refining steps. First it is converted to a liquid by melting, and then it is solidified into bar molds. At this stage it is called *pig iron*. The pig iron bars are the basic ingredient in making some metal alloys, including steel.

A *metal alloy* is made when two or more metals are combined to form a product that is different from any of the materials in their pure form.

The pig iron is purified by burning off—oxidizing—various impurities such as carbon. *Oxidizing* is a process of heating a material to such a high temperature that it actually combines with oxygen and is burned or consumed just as a piece of wood is burned. After the carbon and other impurities have been oxidized out of the pig iron, the iron is ready to be mixed with different materials in controlled amounts. For example, small amounts of carbon are added to create steel.

The steel is created in different types of furnaces using different kinds of fuels. There are Bessemer furnaces, oxygen furnaces, open-hearth electric furnaces, and others.

Other metals also will be discussed in this chapter, but the major focus will be on iron and steel. Iron and steel are called *ferrous metals*, and they are the metals from which most of the machinery and equipment in agricultural industry are made (see Figure 10-1).

Figure 10-1. Iron and steel are used in making agricultural machinery.

CHAPTER GOALS

In this chapter your goals are:

- To identify different kinds of metals you will be working with in agricultural industry
- To demonstrate the methods used to identify metals and the correct metal-working operations for each metal

Ferrous Metals

Ferrous metals are those that contain iron. The iron may be mixed with a small amount of carbon and other materials to produce the desired alloy.

Cast Iron

Cast iron, sometimes called *gray* or *white cast iron*, has high percentages of carbon (from 1.7 to 4 percent), and the carbon is alloyed with the iron as in steel. The carbon is in graphite form. Carbon content in iron is the one factor that determines the kind of iron or steel produced and how it is used.

Because of its high carbon content, cast iron is not easily shaped or bent when cold. In other words, it is not malleable. Cast iron has a very coarse-grained appearance when broken into pieces and carefully examined. When cast iron is broken or fractured, the inside color may be either gray or white, depending on the rate of cooling after it has been heated. If cast iron cools slowly, more graphite is produced, which gives the metal a gray look.

Many parts in equipment and machinery used by agricultural industry, such as engine blocks and heads, are made from gray cast iron. The metal can be drilled, sawed, welded, and machined. Usually, the more ways a metal can be used, the more valuable it is to the industry. However, gray cast iron cannot be bent when it is cold.

As its name implies, cast iron is formed by pouring molten pig iron into desired molds, or casts. This process allows many parts to be made from the same mold (cast) at minimum cost. Because they come from the same cast, the parts generally need no further fitting or machining before they are used.

Malleable Cast Iron

It is possible to make cast iron malleable—able to be shaped when cold—by heating it for a long time. The carbon content in the outer layer of the casting is separated out of the iron. This malleable outer layer allows the casting to be bent slightly without breaking. It is also more resistant to shock and breakage than regular cast iron. Malleable castings are used in tractors for axles, steering arms, levers, hydraulic arms, and other parts. They can be sawed and welded like cast iron.

Today, there is a great quantity of ductile cast iron being produced that is somewhat different from both gray and malleable cast iron because it has greater strength than either one.

Wrought Iron

This is practically pure iron. Almost all the carbon and other impurities are removed in the furnace. Because it is almost pure iron, wrought iron is very malleable. It can be easily bent, shaped, welded, drilled, sawed, and filed. It also resists rust. Its granular structure is fibrous and stringy rather than grainy as in cast iron. With only a trace of carbon, wrought iron cannot be hardened or *tempered*—toughened by first heating and then cooling. Wrought iron is used in decorative iron work as well as for keys and cotter pins.

Mild Steel

This form of iron, *mild steel*, could be classified as being between malleable cast iron and wrought iron. It contains from 0.25 to 0.60 percent carbon, and the carbon is combined with iron to form an alloy. While there is not enough carbon to permit the steel to be hardened or tempered, it is not quite as malleable as wrought iron.

Mild steel is the most common kind of steel. It is used to make and repair many pieces of farm machinery. Like malleable cast iron, it can be drilled, sawed, welded, bent, shaped, ground, and threaded. Mild steel comes in many shapes: angles, bars, sheets, rods, flats, channels, tees, beams, and variations of these (see Figure 10-2).

Tool Steel

This kind of steel contains from 0.60 to 1.50 percent carbon. This is enough carbon to enable the steel to be tempered and hardened. Therefore, as the name implies, tool steel is used to manufacture chisels, punches, shears, and other cutting tools. Tool steel is often classified and sold according to "points." For example, 100-point steel contains 1 percent carbon, and 150-point steel contains 1.5 percent carbon.

Stainless Steel

This is an alloy steel containing varying amounts of chromium and nickel. These ingredients are mixed with the iron in the furnace. Stainless steel is very resistant to rust and the gradual eating away of the metal caused by elements in the air (corrosion). It is used in food-processing equipment, bulk milk storage tanks, and chemical equipment.

Stainless steel is not very malleable and requires specific techniques for shaping and welding. It usually comes in sheet form.

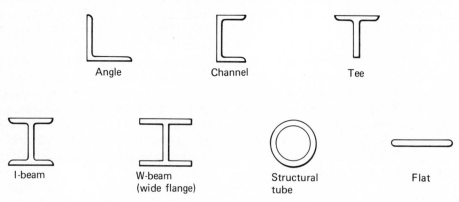

Figure 10-2.   Common shapes of mild steel.

Galvanized Steel

This is usually mild steel that has been "galvanized" by coating it with a thin layer of zinc. The layer of zinc makes the steel resistant to rust and corrosion. Galvanized steel may come in rolls. Often it is used in the construction of outdoor sheds, roofs of farm buildings, and large warehouses because of its ability to resist rust and corrosion. The galvanizing process can be done in two ways—by depositing the zinc electrically or by dipping the steel into molten zinc. Galvanized steel can be worked the same way as mild steel. However, certain precautions must be taken when galvanized steel is welded. When heated, the zinc produces poisonous fumes.

Standard Metal Shapes

Metal is shaped at the foundry through any of five different kinds of processes: *casting, rolling, forging, extruding,* and *drawing.* Depending on the process, the metal may come out in flat sheets or in round bars, flat bars, and bars of other shapes (see Figure 10-2).

Nonferrous Metals

Nonferrous metals contain little if any iron. Among the nonferrous metals are copper, brass, aluminum, lead, and zinc. Usually, these metals can be identified by their color and weight.

Copper

Copper is mined and refined from copper ore. In its pure form, copper is a very soft metal. It melts at 1980°F [1082°C]. Copper is easily identifiable by its rust-red color. Copper is very malleable and can be bent, drilled, sawed, and welded. Because of its softness, however, it cannot be easily ground or filed.

In both its pure and alloy forms, copper has many uses in agricultural industry. Two major uses are for pipes used in plumbing systems and electrical wiring in buildings and motors. Copper is used also for such accessories as starters on tractors because it is an excellent electrical conductor.

Brass and Bronze

Brass and bronze are both alloys created by combining copper with other metals. In the case of brass, copper is combined with from 10 to 40 percent zinc. Bronze is the result of combining copper with zinc and about 10 percent tin. Bronze and brass are soft, malleable metals that are worked in much the same way as copper. Both alloys melt at approximately 1600°F [871°C]. Brass is colored much like copper, while bronze is more reddish-yellow.

Brass and bronze are used often for oxyacetylene brazing rods and for bearings and bushings on farm and ranch equipment. Also, some special-purpose nails and screws are made of brass.

Aluminum

Like iron and copper, aluminum is manufactured from mined ore. In its purest form, aluminum is very soft and malleable. Its strength can be increased by the addition of silicon, manganese, or magnesium. However, when it is formed into an alloy with these other metals, the aluminum becomes less *ductile*—less able to be worked easily.

Aluminum, which is white in color, is very lightweight, resists corrosion, and is a good conductor of electricity (as used in electrical wiring materials). Also, it is used as a substitute for steel where weight is an important factor. For example, certain tools, farm equipment, truck bodies, metal containers, and metal roofing might be made of aluminum for this reason. And aluminum might be used on roof gutters and elsewhere outside because of its resistance to rust and corrosion.

Aluminum can be cut, welded, and drilled, but it is not easily ground or filed because of its softness. It melts at 1218°F [659°C]. When cast in a mold, aluminum can be brittle. Cast aluminum is used for gas engine parts and much of the lawn and garden equipment used by landscaping firms.

Lead and Tin

In its pure form lead doesn't have many uses in agricultural industry. It is used in batteries, however, and lead alloys are used as solder. Common solder is made of 50 percent lead and 50 percent tin. The melting temperature of lead is 621°F [327°C] and of tin is 450°F [232°C], but the melting point of solder is lower than that of either lead or tin. *Solder* is used to join metal materials (see Figure 10-3). For example, solder is used to join copper pipes or tubing. Also it is used for electrical connections.

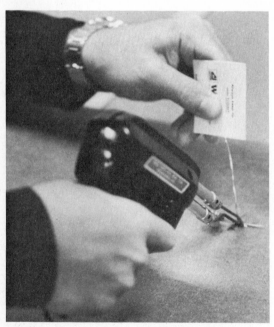

Figure 10-3. A soldering gun.

Testing Materials

Many workers in agricultural industry need to know how to properly identify metals. For example, the mechanic in charge of repairing machinery and equipment needs to know the kind of metal being worked on so that the proper repair or rebuilding can be performed. Persons working on farms and ranches and in agricultural supplies/services need to know about metals if they are going to be able to properly care for equipment or advise on the care of equipment.

Often metals can be identified by their color, weight, and the temperature at which they melt. But, when the worker is uncertain about metals, there are other tests that can be performed in the shop. These methods include the following: spark test, hammer (or ring) test, magnetic test, fracture test, and molten puddle test. In some cases, it may be necessary to perform several tests before proper identification can be made.

Spark Test

The spark test is the most common method of identifying the kind of iron and steel. The spark test is performed by using a power grinding wheel to produce a spark from the metal. The spark test is less effective with nonferrous metals because they are generally softer metals that do not produce the same friction as ferrous metals do when ground.

To produce a spark, the metal is touched lightly to the revolving grinding wheel. The friction caused when the metal touches the wheel burns off (oxidizes) particles of the metal. This burning (oxidation) occurs at different rates for different ferrous metals. The oxidation produces colored sparks, the colors varying with the kinds of metal.

The conditions and characteristics to observe in making the spark test are these:

spark color, spark length, the amount of explosions or bursts of sparks, and the shape of the exploding sparks (such as whether they form a fork).

Wrought Iron. When wrought iron is ground, the sparks given off are straw-colored. They come off the grinding wheel in a long stream, with few explosions or bursts.

Mild Steel. Mild steel produces a series of bright sparks. There are slightly more explosions than with wrought iron.

Tool Steel. A great many bright sparks are produced by tool steel. They come in frequent, short bursts.

Gray and White Cast Iron. Short and frequent bursts of red sparks are given off by both gray and white cast iron.

Malleable Cast Iron. Malleable cast iron produces more red sparks than gray or white cast iron, but the volume is not as great as that for tool steel. Again, the sparks are very explosive.

Alloys. When iron and steel are formed into alloys with other metals—such as magnesium, chromium, nickel, and tungsten—they produce different kinds of sparks. Figure 10-4 shows the kinds of sparks given off by different alloys.

Hammer, or Ring, Test

The hammer, or ring, test is performed by striking the metal a sharp blow with a hammer and listening to the sound (ring) made when the metal is hit. The best way to perfect this test is to practice on known metals. For example, take a piece of cast iron and mild steel and strike each with a hammer. Listen to the ring. Keep doing this until the ring of cast iron can easily be distinguished from the ring of mild steel. The same kind of practice should be tried with steel alloys and nonferrous materials.

Magnetic Test

Ferrous metals, except some forms of stainless steel, are attracted by or cling to a simple magnet. However, nonferrous materials will not cling to a magnet.

Fracture Test

With practice, it is possible to identify metals by looking at the grain when the metal is broken open. For example, cast iron normally has a very coarse grain.

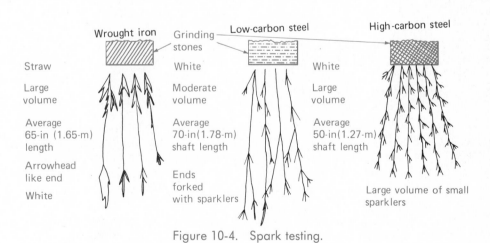

Figure 10-4. Spark testing.

Molten Puddle, or Torch, Test

An oxyacetylene torch (described in Chapter 19) can be used to melt a small portion of a metal. When the metal is melted into a puddle, it should have a shiny, almost mirrorlike appearance. If the puddle is rough, porous, or dull in appearance, usually the metal is an alloy or contains impurities. Iron and iron alloys turn red or orange when heated. When heated, aluminum does not change color, but merely turns to a liquid.

Safety Rules

These safety rules should be followed when metals are being identified.

1. Always wear gloves and safety glasses when grinding for the spark test and while melting metal with an oxyacetylene torch.
2. Know how to operate the power grinder and the oxyacetylene torch before testing materials.
3. Make sure that the metal being tested with the hammer (ring) test is secured so that it won't fly off when struck.
4. Cool metal pieces that have been heated before disposing of them.
5. Handle hot metal with pliers or tongs to avoid being burned.
6. Advise others in the shop when there is hot metal in the area.

Identifying and Selecting Metals: A Review

Agricultural industry—in fact all modern society—relies heavily on metal. If metal had not been discovered and its uses defined, people might still be struggling as did the people of the Stone Age.

Iron is mined, then converted to a liquid, and later combined with carbon and other metals to produce many types of alloys for special applications to agricultural industry—from simple tools, equipment, and machine parts to siding for buildings.

Metals are divided into two kinds: those that contain iron and are called ferrous metals, and those that don't contain iron and are called nonferrous metals.

Metals can be identified by testing them in different ways. If the metal being used is not known to the worker, it must be identified before the worker repairs, saws, tries to shape, or otherwise uses it. Otherwise, the worker may use the metal improperly and damage the project being worked on.

THINKING IT THROUGH

1. Describe the characteristics of the metals in the ferrous group.
2. Describe the characteristics of the metals in the nonferrous group.
3. Describe which test you would use to identify (a) mild steel, (b) aluminum, (c) malleable cast iron, and (d) stainless steel.
4. Describe the spark pattern for (a) mild steel, (b) tool steel, and (c) cast iron.
5. Describe why the spark test is not effective with nonferrous metals.
6. Explain why some ferrous metals are harder than others.

CHAPTER 11

METAL-WORKING TOOLS

The worker in production agriculture, in agricultural mechanics, in agricultural supplies/services, or in other agricultural career fields who knows how to cut, drill, or shape metal is sought after by employers. Most of the equipment and machinery used in agricultural industry today is made of metal. In addition to equipment and machinery, metal is used for many other things.

For example, often metal is used to side a farm or ranch building or to build a fence. Receptacles that hold water, fertilizers, chemicals, and engine fuels often are made of metal.

Just as most agricultural workers need to know how to work with wood, so do most workers need to know how to work with metal.

This chapter will describe tools that cut and drill cold metal and tell how to use them. The person who develops skills with metal-working tools will be valuable in an agricultural job and around the home.

CHAPTER GOALS

In this chapter your goals are:

- To describe the different ways to treat and shape cold metal (working with hot metal is described in Chapter 14)

- To identify the tools best suited for certain jobs and be able to use these tools properly and safely
- To know how to repair metal objects

Metal-Cutting Files

Files are hand-held tools made of hardened steel that use many small chisellike teeth to shape and smooth metal. Files are used to remove small quantities of metal. There are many styles, sizes, and shapes of files. But all files fall into two general categories—single-cut and double-cut. The *single-cut file* has only one set of teeth running across the face of the file. The *double-cut file*, as the name implies, has two sets of teeth, one running at an angle to the other. Figure 11-1 shows the difference between a single-cut and a double-cut file.

Identifying Different Types of Files

The size of a file is determined by its length (exclusive of the tang). The shorter a file is, the finer its teeth are. A large, long file will have coarse teeth for rough filing and removing larger quantities of metal. A 12-in [304.8-mm] file will have coarser teeth than a 6-in [152.4-mm] file. The finer the teeth, the smoother the metal will be when the filing is completed. A file with a rough or

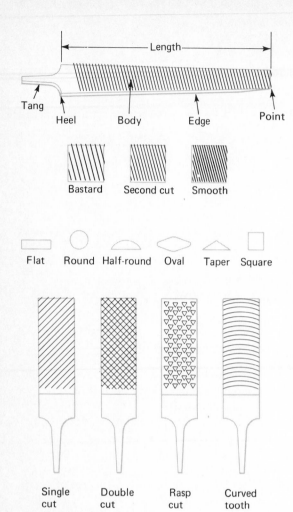

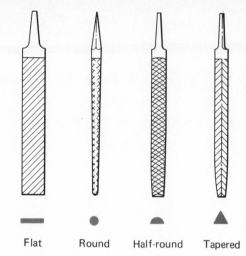

Figure 11-2.   Commonly used files.

Figure 11-1.   Files are selected by length, coarseness, shape, and teeth.

Handles Are Needed

Before any file is used, a handle should be put over the tang. Many persons have been injured while using a file simply because they did not have a handle for the sharp tang. When not in use, files should be hung separately in a rack (see Figure 11-3). The teeth on a file become dull quickly if the file is carelessly stored by throwing it into a drawer or box where it comes in contact with other metal objects.

How to Use Files

When a file is used, it should be held with both hands. The person who is right-handed grips the handle with the right hand and the tip of the file with the left hand. The left-handed person would place the hands in reverse order.

More is accomplished with long, even strokes of the file than with short, rapid strokes (see Figure 11-3). Pressure is applied only on the forward stroke of the file. Light pressure is usually sufficient when the right file is being used for the job. As Figure 11-3 shows, if possible, the ma-

coarse cut is called a *bastard file* (see Figure 11-1).

Files come in such shapes as flat, square, round, half round, and tapered (see Figure 11-2). Files also are described according to the way they are used. For example, the *mill file* is used to remove small amounts of metal. The *auger bit file* is used to sharpen bits so that they bore a smoother hole. The mechanic uses a *thread file* to repair, or redress, damaged bolt threads.

METAL WORKING

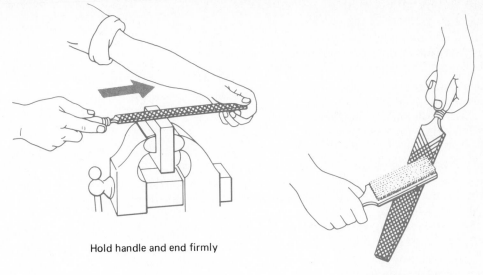

Hold handle and end firmly

Keep file clean.

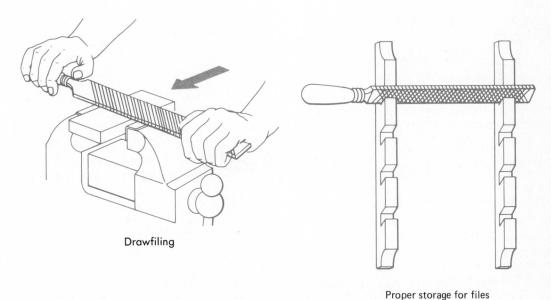

Drawfiling

Proper storage for files

Figure 11-3. Proper file use and care.

terial being filed should be securely fastened by a vise or clamps. A machinist's vise is ideal for holding parts for filing. When a vise is used, the part of the material being worked on should not extend very far above or beyond the jaws of the vise. If the material projects too far from the jaws, the file may cause the metal to vibrate, resulting in poor-quality workmanship.

When heavy filing is being done, or

when soft metals are being filed, the teeth of the file may become clogged with small particles of metal. Usually, these filings can be removed by simply tapping the file sharply against the bench or other hard surface. If this doesn't work, however, it will be necessary to clean the file with a brush called a *file card*. The file card has short metal teeth on one side and tough bristles on the other.

Sometimes a mechanic will need to file a narrow metal strip, rod, or tube. When this is necessary, a different filing stroke is used. This stroke is called *drawfiling* (see Figure 11-3). Instead of filing across the metal at only a slight angle, as would be the case normally, the file is held at a right angle to the metal. Usually a single-cut mill file is used for drawfiling.

Hacksaw

The hacksaw is designed for cutting metal. It consists of a handle, frame, and a blade which is inserted into the frame and can be removed when it is dull or damaged. An important consideration about a hacksaw is the number of teeth per inch in the blade. The number of teeth per inch determines the thickness of the metal and how easily the saw will cut it. Standard blades are 10 or 12 inches [25 or 30 cm]. They are available with 14, 18, or 32 teeth per inch. As a

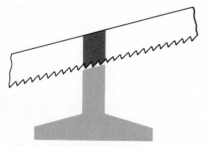

Figure 11-4.   Three teeth must contact the metal to avoid breaking the blade.

Too coarse   Too fine   OK

Figure 11-5.   Select a hacksaw blade with the correct number of teeth for the metal being cut.

general rule, at least three teeth should be in contact with the metal at all times (see Figure 11-4). If only one or two teeth are in contact with the work, the blade is probably too coarse (see Figure 11-5). If 10 or more teeth are in contact with the work, a blade that is too fine probably is being used (see Figure 11-5). Another thing to remember about the blade is that those blades with fewer teeth (coarse-cut blades) may break more easily than those with more teeth (fine-cut ones).

Selecting and Securing the Blade

Hacksaw blades are made of different qualities. The highest quality blade is made from an alloy containing steel and tungsten or molybdenum. Other blades are made of high carbon steel. Naturally, the blade of highest quality also is the most expensive. For most work, the expensive high-speed blade is not really necessary.

After the correct blade has been selected, it should be inserted into the frame so that the teeth are pointing away from the handle. This is done because the cutting is on the forward stroke only. The blade is made secure in the frame by tightening the wing nut below the handle (see Figure 11-6).

How to Use the Hacksaw

The metal being cut should be held securely by a vise or clamps. The point

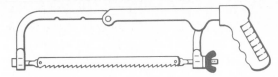

Figure 11-6. Hacksaw cuts should be made on the forward stroke only.

where the cut, or kerf, will be made should protrude only slightly beyond the jaws of the vise. This will help prevent the blade from vibrating and possibly breaking. If the metal being cut is very thin, it can be supported by clamping a wood strip on either side of it. The cut should be made by applying pressure on the forward stroke only. Steady strokes of 40 to 60 per minute should be sufficient for most jobs.

If a cut is not completed because the blade breaks or because the work must be abandoned temporarily, the cut should not be completed with a new blade. The teeth on the new blade will have a wider set and will be damaged by the narrow kerf made by the old blade. The cut can be completed by using another old blade or by sawing with a new blade from the opposite edge of the metal (see Figure 11-7).

Band Saw

The band saw was described in Chapter 8 as it is used in woodworking. The band saw blade used for metal cutting is more fine-toothed than the blade used for cutting wood. Many of the same principles apply to band saw blades as to hacksaw blades. A key factor is the number of teeth per inch. The blades most commonly used for agricultural jobs are those with 10, 14, and 18 teeth per inch [2.5 cm].

How to Use the Band Saw

The metal to be cut should be fastened securely to the cutting table. Otherwise, even a slight movement of the metal can cause the blade to break. The metal should be pushed into the blade slowly and gently to avoid blade breakage. Also, the metal-cutting blade is run at a slower speed than the wood-cutting blade. Set the blade according to the operator's manual or instructor's directions.

The horizontal band saw (see Figure 11-8) should be used only after proper instruction and demonstration. It is best to have the metal positioned so that a

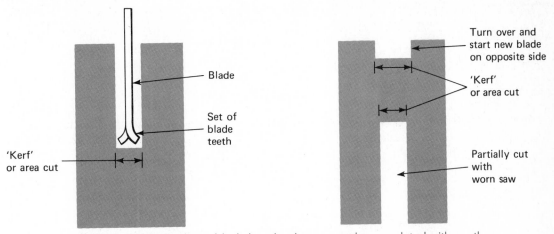

Figure 11-7. If the hacksaw blade breaks, the cut must be completed with another old blade or by sawing with a new blade from the opposite edge of the metal.

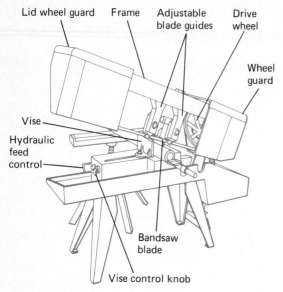

Figure 11-8. Horizontal metal cutting bandsaw.

type of saw or blade is referred to as a "wet-cut."

Long pieces of metal should be supported so that the portion being cut off does not fall and damage the blade, the saw, or the operator. The blade should come into contact with the metal slowly to avoid catching the metal and possibly damaging the teeth of the blade (see Figure 11-9).

Metal-cutting saw blades are available with a raker, wavy, or hook tooth design. The raker tooth is the most common for general use. Also, a blade is designated as either dry or wet cutting. A dry cutting blade should not be lubricated. The wet cutting blade uses a water-soluble oil to cool and lubricate the teeth during the cut. The wet cutting blade will have a much longer life.

minimum of three teeth come in contact with the metal. This means, for example, that if the blade has 14 teeth per inch, then a ¼ in [6.3 mm] or thicker piece should be clamped on the edge. A piece less than ¼ in [6.3 mm] should be clamped in the flat position. If possible, angle iron and channel iron should be placed in the vise so that the saw blade cuts both legs at the same time. Under certain circumstances (where the kind of metal and/or the blade dictates a difficult cutting operation), the blade and the metal may be slightly lubricated. This

Saber Saw

The saber saw, which is also used in woodworking, was discussed in Chapter 8. This electrically operated saw can be used for cutting metal by using a steel cutting blade. The saber saw has a particular value for cutting up fuel barrels and water heater tanks. It is not safe to cut fuel tanks with a cutting torch because of the danger of explosion. If the saber saw has variable or dual speeds, the slowest speed should be used for cutting metal.

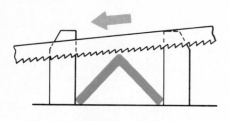

Cutting angle iron

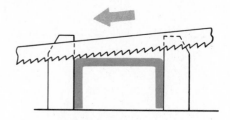

Cutting channel iron

Figure 11-9. The efficiency of the horizontal bandsaw can be improved by correctly positioning the metal.

METAL WORKING

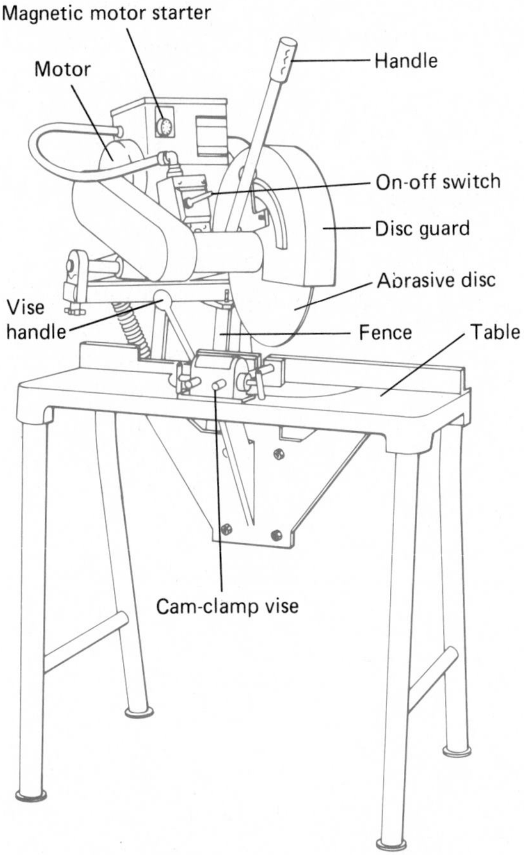

Abrasive Cutoff Saw

The abrasive cutoff saw is a relatively new electrical tool, but one that is growing in popularity among agricultural workers. It is a fast and accurate method of cutting metal. The saw is a rather simple piece of equipment (see Figure 11-10). The cutting action is done with a reinforced abrasive blade. Reinforced blades are held together with strong fibers that prevent them from shattering while in use.

Before the cutoff saw is used, the operator should put on gloves, protective goggles, and long-sleeved clothing. Sparks and particles of metal are produced by the high-speed saw blade. The cutoff saw should not be used near chemicals or explosive materials because of the sparks. The operator should stand to one side when using the saw to avoid being struck by sparks and metal bits.

It is important to properly position the metal to be cut. For example, a heavy, flat piece of metal should be clamped to the table in the vertical position instead of horizontally (see Figure 11-11). Metal with angular shapes should be placed so that a minimum arc of contact with the blade is maintained (see Figure 11-11).

Drill Press

The drill press was also described in Chapter 8 in connection with woodworking, and

Figure 11-10. Abrasive cutoff saw.

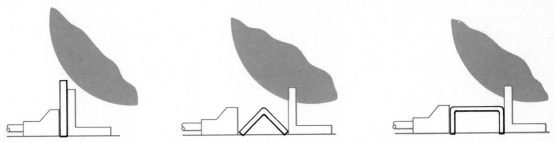

Figure 11-11. Positioning metal in the abrasive cutoff saw to improve cutting speed and blade life.

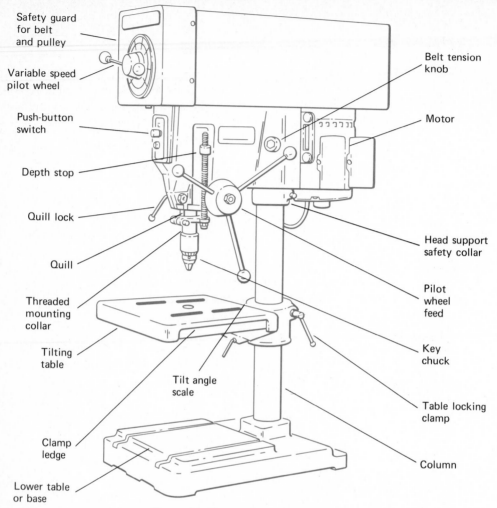

Safety guard
for belt
and pulley

Variable speed
pilot wheel

Push-button
switch

Depth stop

Quill lock

Quill

Threaded
mounting
collar

Tilting
table

Tilt angle
scale

Clamp
ledge

Lower table
or base

Belt tension
knob

Motor

Head support
safety collar

Pilot
wheel
feed

Key
chuck

Table locking
clamp

Column

Figure 11-12.   Pedestal drill press.

it is discussed in Chapter 12 in connection with preparing metal for tapping and threading. The parts of the drill press, as shown in Figure 11-12, should be studied carefully. The drill press has either a multiple-step belt pulley or a variable speed adjustment. The speed should be determined according to the operator's manual or the directions of the instructor. The larger the diameter of the hole to be drilled and the harder the metal, the slower the twist drill should run. (See Table 12-3.)

How to Use the Drill Press

Before drilling is started, the hole should be located with a center punch. This will make it easier for the drill to start. After the center punch has been used, the metal should be secured to the table with vises or C-clamps furnished with the press (see Figure 11-13). After selecting the proper twist drill, insert the drill into the chuck. Check the operator's manual for instructions concerning lubrication of the drill and

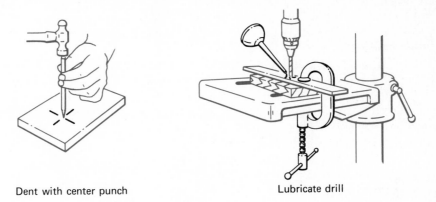

Dent with center punch

Lubricate drill

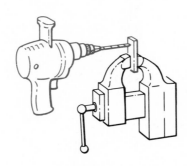

Hold metal in vise
when hand drilling

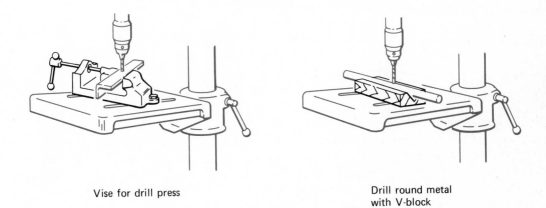

Vise for drill press

Drill round metal
with V-block

Figure 11-13.   Recommended procedures for drilling metal.

the metal. Firm pressure is applied on the press handle in drilling. If holes larger than ⅜ in [9.5 mm] are to be drilled, it will help to first drill a smaller pilot hole.

When the drill is almost through the metal, the pressure is reduced. Otherwise, the drill may catch the metal and jerk it loose from the vise or C-clamps. If this happens, the metal may begin to spin along with the drill. Spinning metal is dangerous to the operator, and if it spins off the drill, it can be thrown across the room and injure someone. It is necessary to maintain correct pressure on the feed handle, to hold the metal on the table and prevent it from being thrown from the drill. While the pressure is applied on the handle, the power switch should be turned off. When the drill stops turning, release the pressure on the handle.

Selecting Twist Drills

There are many sizes of twist drill to choose from. Figure 11-14 also describes the parts of a twist drill: the shank, body, cutting lips, and point. The size of twist drills is determined and sold by four different designations: *alphabetical, wire-gauge size, fractional,* and *metric*. Fractional is the most common designation used. Fractional sizes start with ¹/₆₄ in and increase by ¹/₆₄ in up to 1¾ in, then by ¹/₃₂ in up to 2¼ in, and then by ¹/₁₆ in up to 3½ in.

The alphabetical sizing starts with A at 0.234 in. The sizes then increase up to 0.413 in (Z). Wire-gauge twist drills are numbered from 80 to 1, with 80 being 0.0138 in and 1 being 0.228 in. These twist drills are used mostly by machinists.

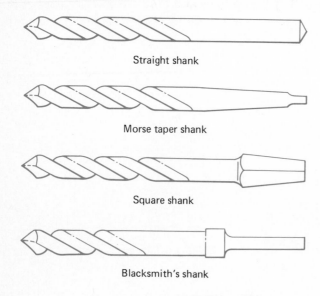

Straight shank

Morse taper shank

Square shank

Blacksmith's shank

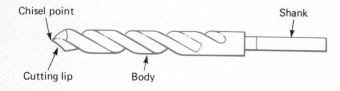

Chisel point

Shank

Cutting lip

Body

Figure 11-14. Four common twist drill shanks and parts.

Twist drills that are sized according to the metric system start at 0.35 mm and increase by 0.05 mm. A twist drill of 0.35 mm is equivalent to an 80-wire-gauge twist drill and is slightly smaller than a $\frac{1}{16}$-in twist drill.

Metal Shear

Light-gauge sheet metal usually is cut best by using a metal-cutting shear. The shear is less expensive than many saws, and it does the job quickly.

Most shears have two cutting blades, one stationary and the other movable. The blades are made of high-carbon, high-chrome steel (see Figure 11-15). Each make and model has a plate that indicates the size or capacity of the shear. Metal that is larger than the capacity of the shear should not be cut or the blades will be damaged.

Box and Pan Brake

It is possible to make a tool box, oil drip pan, and many other useful items for shop and home by using a box and pan brake (see Figure 11-16). The machine has removable parts called *fingers* that clamp and hold thin-gauge sheet metal in different positions for bending and shaping.

Once the desired number and width of the fingers are determined (the operator's manual or the instructor can help), the metal is clamped beneath the fingers and the sheet metal is bent or formed by lifting the folding blade to the desired angle of bend.

All box and pan brakes have a plate attached that indicates the capacity of the machine. Metal of a heavier gauge than called for by the directions should not be used. The machine is equipped with finger gauges and other stops that make it possible to duplicate parts.

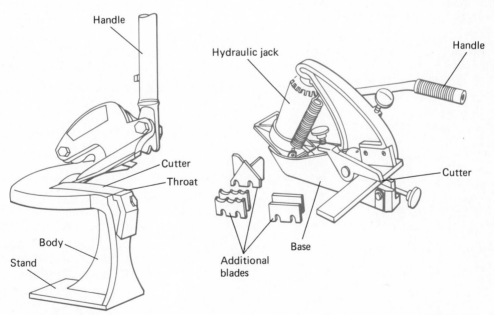

Figure 11-15. Two types of metal shears used to cut sheet metal (left) and flats or shapes (right).

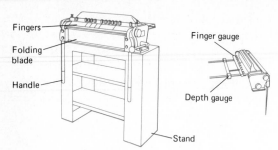

Fingers

Folding blade

Handle

Finger gauge

Depth gauge

Stand

Figure 11-16. The box and pan break machine is used to shape sheet metal.

Slip Roll (Forming) Machine

The slip roll, or forming, machine is used to form curved sheet metal and to form cylinders (see Figure 11-17). The machine has three rollers, two of which are the front pinch rollers that grip the metal and force it against the rear roller for shaping.

The two front pinch rollers are adjusted so that they pull the metal into the machine. The rear roller also is adjustable, according to the kind of metal being shaped. Steel, for example, will have more spring than softer metal. The way to learn how to adjust the rear roller is through experimentation with different kinds and thicknesses of metal.

The slip roll machine also has a plate attached that specifies its capacity. Metal

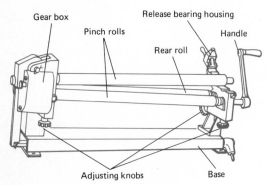

Gear box

Pinch rolls

Release bearing housing

Handle

Rear roll

Adjusting knobs

Base

Figure 11-17. The slip roll machine is used to form curved sheet metal and cylinders.

that is thicker than the rated capacity of the machine should not be formed. Otherwise, the housings holding the ends of the rollers will break.

In forming a cylinder, it is better to make several passes through the machine with the metal than to try to do it in only one pass.

Safety Rules

These safety rules should be followed when you are using metal-working tools:

1. Make sure files are in good condition and are properly fitted with a handle.
2. Use a file only for the purpose intended. It should not be used as a pry bar or hammer.
3. When you use the band saw, make sure that the wheel guards are in position and that the blade is in good condition.
4. To avoid blade damage make sure the blade is adjusted so there is the proper tension.
5. Long pieces of metal should be supported so that cut pieces do not fall from the saw.
6. Goggles and protective clothing (long sleeves and gloves) should be worn when a revolving machine such as a cutoff saw is used. Loose-fitting clothing should not be worn.
7. Jewelry that may get caught in the machine should not be worn, and long hair should be kept under a net or cap.
8. Never stand directly in front of a turning blade; always stand to the side.
9. To avoid a possible explosion, a saber saw with metal-cutting blade should be used for cutting up barrels that may have contained fuel or other flammable material. These containers should be filled with water before cutting.

METAL WORKING

Metal-Working Tools: A Review

All metal products have been cut, bent, drilled, smoothed, fastened, or changed by some process. The agricultural worker who is familiar with these processes is a valuable employee in the mechanic's shop, on the farm or ranch, or in the firm selling and servicing equipment and chemicals.

Tools can be dangerous when used improperly or carelessly. It is most important, therefore, to be very familiar with the tools and to observe recommended safety rules.

THINKING IT THROUGH

1. What are the types of hand files which are commonly used in agricultural industry?
2. Describe why it is important to have handles on all files.
3. Describe the correct use of the hand hacksaw.
4. What is the key to selecting the coarseness of the saw blade?
5. Sketch how metal shapes should be positioned in the metal bandsaw.
6. What are the four designations used on twist drills?
7. Why should metal always be placed in a vise before drilling?

CHAPTER 12

DRILLING, TAPPING, AND THREADING

The purpose for cutting and shaping metal may be to prepare two or more pieces to be joined in some way. The mechanic or worker in an agribusiness store may want to build a metal tool cabinet. The employee at a processing plant may want to repair or rebuild a piece of equipment or machinery.

To assemble or construct machinery or make the repairs, it is necessary to join pieces of metal with bolts. Holes have to be drilled in the metal and prepared to receive the bolts. And the bolts must be prepared to fit into the holes. These two processes are called *tapping* and *threading*.

Many agricultural workers have to do tapping and threading occasionally; some must do it frequently. Six different kinds of workers were asked whether they had to tap and thread and know the skills that go along with these processes. The results of the survey are given in Table 12-1.

Tapping and threading require precise work and the use of rather expensive tools. Therefore, only those persons properly trained will be able to do tapping and threading on the job. Employers can't afford costly mistakes.

CHAPTER GOALS

In this chapter your goals are:

- To demonstrate how to prepare for drilling, tapping, and threading
- To correctly use layout and measurement techniques and tools
- To demonstrate the procedure used to thread the inside of a drilled hole and to thread a bolt
- To remove bolts that have broken off and cannot be removed with a wrench

Layout and Measurement

If the cabinet is to be constructed properly and the repairs are to be made correctly, it is necessary to know precisely what to do and how to do it. That's where layout and measurement come in. First carefully review the shop plan. This may be an exact blueprint with specified measurements and instructions or a rough sketch showing just basic ideas and approximate design.

In any case, it is necessary to determine where machining is required. Marking the locations is called *layout*. It may be done with simple tools or with rather sophisticated instruments. The selection of tools depends on the job requirements. For the most part, simpler tools and techniques will be used. Many of the layout and measurement tools have been described in Chapter 6. The basic tools required in

TABLE 12-1.

	FARMER & RANCHER	SUPPLIES WORKER	TRACTOR MECHANIC	PACKING-HOUSE WORKER	GREEN-HOUSE WORKER	LOGGER
Using measuring tools	X	X	X	X	X	X
Using layout tools	X		X			
Using twist drills	X	X	X	X		X
Threading bolts	X	X	X	X		X
Threading pipe	X	X	X	X	X	
Tapping a hole	X		X	X		X
Removing broken bolts	X		X	X		X

metal layout are the combination square, scribe, center punch, protractor, dividers, calipers, and a good measuring tape or scale.

In addition to these tools, it is good to have something that can show measurements clearly on the metal. On rough cast iron or castings, liquid white shoe polish can be used to make the marks more visible. Often a blue toolmaker's ink is used on finished steel. Soapstone is good for circling a center punch mark so that it can be located easily. Layout material remaining on the metal after the job is completed can be removed with a cleaning solvent. In marking metal for a gas cutting torch, it is good to mark the line with occasional punch marks, because soapstone, for example, will burn off.

Locating Holes to be Drilled

The most common layout job is determining where the holes should be drilled. The best way to start is to find the center of the metal. To find the center of a rectangular piece, diagonal lines are scribed between the opposite corners. The lines will cross at the exact center (see Figure 12-1). To find the center on the end of round metal, use a combination square equipped with the centering head (see Figure 12-1). Any line crossing the center point of the inside edge of the square's blade will intersect the exact center of the round piece. Reference marks, of course, can be laid out on flat surfaces with the combination square and a good straightedge or measuring tape.

When the marks have been made, the center punch is used. The center punch should make a hole large enough to fit the chisel point of the twist drill (see Figure 12-2). The mark is made by striking the punch with a sharp blow from a ball peen hammer. After the punch is used, it is a good idea to recheck the measurements before drilling is begun. The location might be checked with a tape measure or a ver-

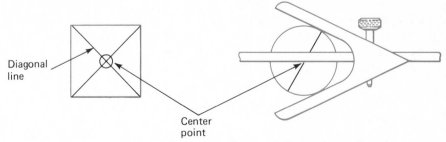

Figure 12-1. Locating the center of rectangular and round stock.

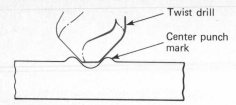

Figure 12-2. The center punch mark ensures accuracy by providing the exact location for starting the twist drill bit.

nier caliper. Intersecting lines can be scribed through the punch mark for greater accuracy.

When the hole is being precision drilled, it is wise to check to see that it is *concentric*. That is, the circumference of the hole should intersect the lines marking the hole at a point that is equally distant from the center of the hole. If the hole is not centered, the center mark can be repunched. However, this must be done very early in the drilling process.

Drilling

Before going further, it may be desirable to quickly review points made about drilling in Chapters 9 and 11. Information about caring for and redressing twist drills is found in Chapter 23.

Most drilling of metal takes place on a drill press, although the hand-held power drill is occasionally used. Whenever the drill press is used, the metal being drilled must be firmly secured to the table (see Figure 12-3). This may be done with a drill press vise, with a strap clamp, or by bolting a pair of stops to the slots in the table. Without something to hold it down, the metal will move on the table. This definitely will cause the work to be inaccurate. It may also cause injury. If not held down, the twist drill may seize the metal and throw it against either the column of the press or into the operator.

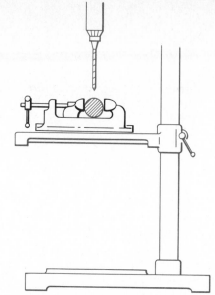

Figure 12-3. All metal must be secured to the table before drilling.

One easy method to prevent flat stock from rotating on the table is to clamp or bolt a stop in the slots of the table and hold the work against the stops (see Figure 12-4). If the twist drill should seize, the stops will prevent the metal from spinning.

To hold small round stock in place, either a V-block with a Y-clamp or a

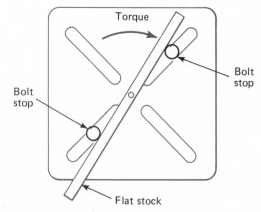

Figure 12-4. Bolt stops will prevent flat stock from turning while drilling.

METAL WORKING

V-block with a step block that has a strap clamp can be used. Round stock presents a problem, because the drill will tend to slide off the punch mark. To avoid this slippage as much as possible, the hole should be drilled through the exact center of the stock. The location can be checked by using a combination square on the table and sighting along the drill and the round stock.

Irregularly shaped pieces are a special problem. One solution is to clamp a heavy metal angle iron to the table and then fasten the metal piece to be drilled to the vertical leg of the angle iron with C-clamps. Still another possibility is to use a tapered piece as a shim, or wedge, to support the irregular metal work in the drill press vise. However it is done, the point is to make sure the metal is as secure and level as possible.

Most drill presses are equipped with both a hole and slots in the table. The hole is to be used with a boring bar. It should *not* be drilled through! If the hole is drilled through, eventually the drill press table itself will be damaged.

Using the Drill Press

The correct speed and pressure for drilling must be determined from experience. The ''feel'' for drilling can be achieved by applying the drill to the metal and watching the chips or cuttings. When mild steel is drilled, for example, the cuttings should spiral from the hole with a uniform thickness. If the cuttings are blue in color, too much pressure is being applied. If the cuttings break or chip into small pieces, either the twist drill is improperly conditioned or the feed—the pressure applied to the drilling—is too slow.

It is very important that the work be both lubricated and cooled with commercial *cutting fluid*. The fluid reduces the friction between the cutting lips and the metal. It also absorbs the heat from the drilling process. All drilling jobs, with the exception of those involving cast iron, require the use of a cutting fluid. Table 12-2 shows what fluids can be used with different metals.

Speed is an important factor in the use of a twist drill. If the speed is too fast, the drill may be damaged and the work may have a poor finish. Table 12-3 gives the maximum recommended speeds in revolutions per minute (rpm) for different-sized twist drills on mild and tool steel.

Types of Bolt and Screw Threads

Imagine coming down a spiral ramp, such as one descending from a theater or stadium. Round and round and down you

TABLE 12-2. Recommended Cutting Fluids for Various Metals

	ALUMINUM	BRASS	CAST IRON	CAST STEEL	MILD COPPER	TOOL STEEL	WROUGHT STEEL	IRON
Kerosene	X							
Kerosene and Lard Oil	X	X						
Soluble Oil	X	X		X	X	X	X	X
Mineral Lard Oil						X	X	X
Sulphurized Oil					X	X	X	X
Lard Oil						X		
Dry			X					

TABLE 12-3.   Maximum Recommended Speeds for Twist Drills in RPM's*

DRILL DIAMETER (in)	(mm)	MILD STEEL	TOOL STEEL	DRILL DIAMETER (in)	(mm)	MILD STEEL	TOOL STEEL
1/16	1.60	4500	3500	3/4	19.00	600	300
3/32	2.40	4000	2675	13/16	20.50	550	280
1/8	3.20	3600	1850	7/8	22.00	525	260
5/32	4.00	3000	1500	15/16	23.50	485	250
3/16	4.75	2450	1200	1	25.50	450	225
1/4	6.30	1825	925	1 1/8	28.50	400	200
5/16	7.90	1500	725	1 1/4	31.50	360	180
3/8	9.50	1200	600	1 3/8	35.00	325	160
7/16	11.00	1050	500	1 1/2	38.00	300	150
1/2	12.50	900	450	1 5/8	42.00	275	140
9/16	14.00	800	400				
5/8	15.50	725	350				
11/16	17.50	650	325				

*This table is not an equivalent conversion; rather, it is designed to show only the recommended twist drill speeds.

go. The thread of a common bolt or screw is much like that spiral ramp. Starting at the top, the thread goes around and down (see Figure 12-5).

Basically five types of threads are used in agricultural industry. They are: national coarse (NC), national fine (NF), international standard organization coarse (ISO coarse), international standard organization fine (ISO fine), and national pipe thread (NPT).

Identifying and Using Different Threads

The NC and NF types are based on the U.S. Customary System of measurement and are most common among cap screws and bolts. The ISO threads use the metric system and are used on all foreign-made machinery. They are becoming increasingly popular in the United States. They may be identified by an ISO marking on the head.

The NC thread is stronger and is less affected by rust and high temperatures. The coarse thread assembles easier and quicker and resists cross threading. Fine threads are used on engine bolts and other bolts that are subject to high torque, or twisting force, and vibration. The fine thread is less likely to loosen under loads and stress.

Because there are different types of threads, it is important to pay attention to the type of thread when a bolt, nut, or screw is replaced. There is a U.S. Customary-metric screw pitch gauge that can help identify the type of thread. The tool has a number of serrated blades that fit into the threads. Each blade is designated by the number of threads per inch or centimeter.

The NC, NF, ISO coarse, and ISO

Figure 12-5.   The bolt advances A distance with each complete revolution.

METAL WORKING

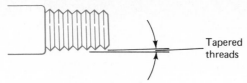

Figure 12-6. Pipe threads taper to ensure a

Adjustable die   Round split die

Figure 12-7. Two types of dies.

fine threads are cut at the same depth throughout their length and are designed to be used in a tapped hole or with a nut.

The NPT threads differ from the cap screw and bolt thread mentioned in that they are tapered. The threads usually taper about $^1/_{16}$ in [1.6 mm] for every linear inch [25.4 mm] of pipe (see Figure 12-6). However, this can vary with the size of the pipe. The taper allows the pipe to self-tighten when it is screwed into a fitting. The NPT threads sometimes are referred to as BSPT (Briggs standard pipe threads) or ISO pipe threads.

Each cap screw or bolt is available in a variety of strengths. The designated grade markings are ASTM (American Society of Testing Materials) and SAE (Society of Automotive Engineers). Table 12-4 lists the grade markings and tensile strengths for common steel bolts.

Applying Bolt Threads

Threads that are cut on a rod or bolt are outside threads and are applied with a hardened steel tool called a *die*. The die is available in either U.S. Customary System or metric thread and may be either one piece (a round split die) or an adjustable die. The round split die is used generally, but the adjustable die allows the mechanic to vary the depth of the thread. The variations allow for a close, medium, or free fit, depending on the application and use of the bolt (see Figure 12-7). The tool used to hold the die is called a *die stock*. The complete

set of threading tools is called a *screw plate set* or a *tap and die set*.

Making the Threads

The threading operation is started by making sure the end of the bolt or rod is square. If necessary, the end can be squared by using a bastard file. After squaring, file a 45° chamfer around the edge of the bolt. With the die inserted into the die stock, the taper of the die should start on the bolt first. This allows all cutting points of the die to remove a small amount of material from the bolt. Even though the die has tapered cutting points, the formed threads will not taper.

The die is held squarely over the bolt, and moderate downward pressure is applied. The die is turned clockwise halfway around the bolt. Adequate cutting fluid is applied to the thread area. Another half turn is then made. This should get the die started straight. The die stock continues to be turned, with cutting fluid applied every two or three turns. It is a good idea to occasionally give a quarter-turn counterclockwise to break the metal chips. This is particularly important in threading a long bolt. The metal chips can be removed with a small screwdriver or a piece of wire. Chips that are not removed will be ground into the threads.

When the desired length has been threaded, a quarter-turn counterclockwise should be made to remove filings. The die

TABLE 12-4.   ASTM and SAE Grade Markings for Steel Bolts

GRADE MARKING	SPECI-FICATION	MATERIAL	BOLT SIZE (in)	PROOF LOAD MIN. (psi)	TENSILE STRENGTH MIN. (psi)
				PHYSICAL PROPERTIES	
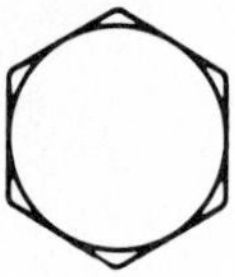	SAE—grade 0	Steel	All sizes	. . . .	. . . .
	SAE—grade 1 ASTM—A 307	Low carbon steel	All sizes	. . . .	55,000
	SAE—grade 2	Low carbon steel	Up to ½"	55,000	69,000
			Over ½" to ¾"	52,000	64,000
			Over ¾" to 1½"	28,000	55,000
	SAE—grade 3	Medium carbon steel, cold-worked	Up to ½"	85,000	110,000
			Over ½" to ⅝"	80,000	100,000
	SAE—grade 5		Up to ¾"	85,000	120,000
			Over ¾" to 1"	78,000	115,000
			Over 1" to 1½"	74,000	105,000
	ASTM—A 325	Medium carbon steel, quenched and tempered	Up to ¾"	85,000	120,000
			Over ¾" to 1"	78,000	115,000
			Over 1" to 1½"	74,000	105,000
			Over 1½" to 3'	55,000	90,000
 BB	ASTM—A 354 grade BB	Low alloy steel, quenched and tempered (medium carbon steel, quenched and tempered may be substituted where possible)	Up to 2½"	80,000	105,000
			Over 2½" to 4"	75,000	100,000
BC	ASTM—A 354 grade BC	Low alloy steel, quenched and tempered (medium carbon) steel, quenched and tempered may be substituted where possible)	Up to 2½"	105,000	125,000
			Over 2½" to 4"	95,000	115,000
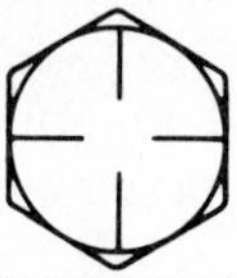	SAE—grade 6	Medium carbon steel, quenched and tempered	Up to ⅝"	110,000	140,000
			Over ⅝" to ¾"	105,000	133,000

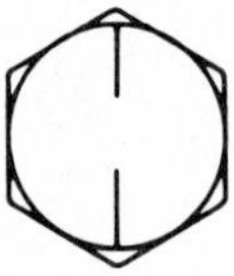

TABLE 12-4.   ASTM and SAE Grade Markings for Steel Bolts (continued)

GRADE MARKING	SPECI-FICATION	MATERIAL	PHYSICAL PROPERTIES		
			BOLT SIZE (in)	PROOF LOAD MIN. (psi)	TENSILE STRENGTH MIN. (psi)
	SAE—grade 7	Medium carbon alloy steel quenched and tempered, roll threaded after heat treatment	Up to 1½"	105,000	133,000
	SAE—grade 8 ASTM—A 354 grade BD	Medium carbon alloy steel, quenched and tempered	Up to 1½"	120,000	150,000

ASTM Specifications

A 307—Low carbon steel externally and internally threaded standard fasteners.

A 325—Quenched and tempered steel bolts and studs with suitable nuts and plain, hardened washers.

A 354—Quenched and tempered alloy steel bolts and studs with suitable nuts.

SAE Specification

Physical requirements for threaded fasteners.

is then turned counterclockwise until it comes off the bolt. The thread should be checked with the appropriate nut. The end of the thread can be redressed with the bastard file to remove any wire edges.

Applying Pipe Threads

The pipe, of course, should be cut to the correct length with a pipe cutter prior to threading. Enough length should be allowed for the pipe to go into the fitting. If the pipe will be used to transport air, feed, or any liquid, the burr must be removed from inside the pipe by using a pipe reamer. A tapered reamer is used to avoid damaging the inside of the pipe. The tapered reamer is inserted into the end of the pipe and turned clockwise.

The pipe must be square before the threads are applied. If it is not square, it can be squared with a bastard file. Then the proper size die is inserted into the die stock so that the guide and the tapered cutting edges start on the pipe first. With the die stock held in the palm of the hand, the handle is turned clockwise half a turn. Cutting fluid is applied to the thread area. The turning of the die should continue, and cutting fluid should be applied after a full thread is completed.

When the threading has been completed, the die should be turned counterclockwise until the die breaks the cuttings loose from the thread. All cuttings are then removed with a small screwdriver or a wire. The threads are inspected. If necessary, the end of the pipe thread can be redressed with a bastard file to remove the wire edge. All excess cutting fluid should be wiped away.

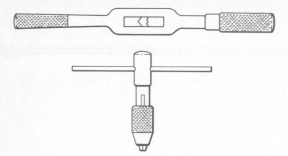

Figure 12-8.   Two types of tap wrenches.

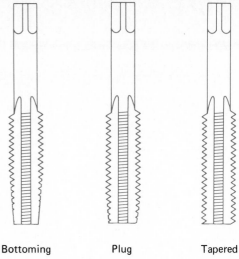

| Bottoming | Plug | Tapered |

Figure 12-9.   Three standard types of taps.

Tapping a Hole

Tapping a hole refers to the application of thread to the inside of a hole. These threads are inside threads; they mate with the outside thread on the bolt or pipe. The threads are applied with a hardened-steel tool called a *tap wrench* (see Figure 12-8). Like dies, taps are available in machine screw and fractional sizes with U.S. Customary System or metric threads. The size of the tap usually is stamped on the shank. For example. a $^3/_8$–16 tap has a $^3/_8$-in diameter with 16 threads per inch. This is classified as a national coarse thread.

Types of Taps

There are three standard types of taps: *tapered*, *plug*, and *bottoming* (see Figure 12-9). The tapered tap is designed for cutting threads in an open hole. It also is used before a plug tap is, because it has a longer cutting surface. The plug tap is designed for applying threads where a self-tightening thread is required, such as a socket

TABLE 12-5.   Recommended Tap and Drill Sizes—English Measurements

BOLT DIAMETER (IN)	NATIONAL COURSE (NC)		NATIONAL FINE (NF)		SCREW EXTRACTOR	
	THREADS PER IN	DRILL SIZE	THREADS PER IN	DRILL SIZE	SCREW NO.	DRILL SIZE
$^1/_4$	20	$^{13}/_{64}$	28	$^7/_{32}$	1	$^5/_{64}$
$^5/_{16}$	18	$^1/_4$	24	$^{17}/_{64}$	2	$^7/_{64}$
$^3/_8$	16	$^5/_{16}$	24	$^{21}/_{64}$	3	$^5/_{32}$
$^7/_{16}$	14	$^{23}/_{64}$	20	$^{25}/_{64}$	3	$^5/_{32}$
$^1/_2$	13	$^{27}/_{64}$	20	$^{29}/_{64}$	4	$^1/_4$
$^9/_{16}$	12	$^{31}/_{64}$	18	$^{33}/_{64}$	4	$^1/_4$
$^5/_8$	11	$^{17}/_{32}$	18	$^{37}/_{64}$	5	$^{17}/_{64}$
$^3/_4$	10	$^{21}/_{32}$	16	$^{11}/_{16}$	5	$^{17}/_{64}$
$^7/_8$	9	$^{49}/_{64}$	14	$^{13}/_{16}$	6	$^{13}/_{32}$
1	8	$^7/_8$	12	$^{15}/_{16}$	6	$^{13}/_{32}$

METAL WORKING

TABLE 12-6. Recommended Tap and Drill Sizes—ISO-Coarse

BOLT DIAMETER (mm)		PITCH	DRILL SIZE
1st CHOICE	2nd CHOICE	mm/Thread	(mm)
1		0.25	0.75
	1.1	0.25	0.85
1.2		0.25	0.95
	1.4	0.3	1.1
1.6		0.35	1.25
	1.8	0.35	1.45
2		0.4	1.6
	2.2	0.45	1.75
2.5		0.45	2.05
3		0.5	2.5
	3.5	0.6	2.9
4		0.7	3.3
	4.5	0.75	3.75
5		0.8	4.2
6		1	5
8		1.25	6.75
10		1.5	8.5
12		1.75	10.25
	14	2	12
16		2	14
	18	2.5	15.5
20		2.5	17.5
	22	2.5	19.5
24		3	21
	27	3	24 or $1^{5}/_{16}''$
30		3.5	26.5 or $1^{3}/_{64}''$
	33	3.5	29.5 or $1^{5}/_{32}''$
36		4	32 or $1^{1}/_{4}''$
	39	4	35 or $1^{3}/_{8}''$
42		4.5	37.5 or $1^{31}/_{32}''$
	45	4.5	40.5 or $1^{19}/_{32}''$
48		5	43 or $1^{11}/_{16}''$

set screw. The plug tap also is used prior to a bottoming tap. The bottoming tap is used in a blind hole, where the hole does not go through and the stud bottoms out inside. The bottoming tap has a very small chamfer, or edge, on the thread and should be used only following the use of both the tapered and plug taps.

To tap a hole, first the correct size twist drill is selected (refer to Tables 12-5 and 12-6). If a $^{3}/_{8}$–16 bolt is to be used, the hole is drilled first with a $^{5}/_{16}$-in twist drill. This allows sufficient base metal for proper depth of the thread in the hole. From Tables 12-5 and 12-6, the correct size tapered tap can be selected and secured into the jaws of the tap handle.

While the tap stock is held in one hand, the tap is inserted squarely into the hole (see Figure 12-10). Moderate down-

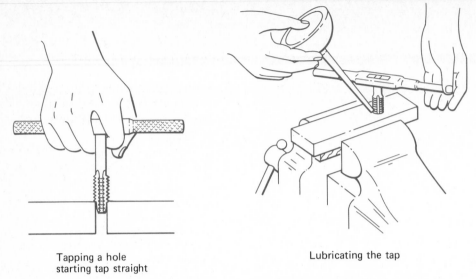

Tapping a hole
starting tap straight

Lubricating the tap

Figure 12-10. Two important aspects of ensuring quality threads.

ward pressure is used, and the tap stock is turned clockwise one full turn. The clockwise turning should be continued, with the correct cutting liquid. As the turning continues, only slight downward pressure should be needed. The pitch of the thread provides the correct feed.

Continue turning until the tap reaches the bottom of the hole or turns free after leaving the hole. If chips clog the tap, it can be turned counterclockwise a quarter-turn to remove the cuttings. When the threading has been completed, the tap is backed out by turning the handle counterclockwise. All excess cutting fluid should be removed.

Removing Broken Bolts

It is not uncommon for an agricultural worker to find that a bolt in a threaded hole in an iron casting has broken off. Sometimes the cause of the problem is that too much torque was applied during the installation. On other occasions, the bolt has corroded or rusted in its hole.

To remove a rusted bolt, first apply a high-quality penetrating oil and allow it to thoroughly soak into the thread cavity. However, if the bolt breaks off, a screw extractor will be needed to remove the broken-off bolt from the hole.

The screw extractor is tapered and has a left-hand spiral for removing bolts with right-hand threads. First, the diameter of the bolt is determined. The correct drill size is identified according to the information given in Table 12-5. The broken bolt should be marked with a center punch exactly in the center. A pilot hole is then drilled with the correct twist drill. Many manufacturers specify the depth of the pilot hole so the extractor will make contact with the bottom of the pilot hole as the spirals dig into the bolt.

By using the same tables, the correct screw-extractor size is selected. Screw extractors are sized by numbers from 1 to 12. A number 1 works on a bolt sized $^3/_{16}$ to $^1/_4$ in [4.8 to 6.4 mm]. The screw extractor is used with a tap stock. The same general procedures as used for tapping are followed. Pressure is applied, and the handle is turned counterclockwise until the broken bolt is removed (see Figure 12-11).

METAL WORKING

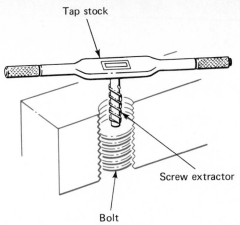

Tap stock

Screw extractor

Bolt

Figure 12-11. A broken bolt is often removed with a screw extractor.

The screw extractor is made of high-carbon steel. Under severe pressure, therefore, it may break. If it does, the broken parts may be removed from the hole by using needle-nose pliers.

If the bolt can't be removed with the screw extractor, another possibility is to drill it out with a twist drill that is slightly smaller than the broken bolt. In doing this, it is important that the drill does not cut into the side of the hole. Then the shell of the bolt can be removed with a prick punch and small cold chisel.

The broken bolt also may be removed by welding a nut to the top of the bolt. This process is described in Chapter 18. Another method is to drill the broken bolt with a slightly oversized drill and to re-thread for a larger-sized bolt. A commercial thread insert, available from most auto parts stores, can also be used.

Safety Rules

These safety rules for drilling, tapping, and threading should be followed:

1. All metal pieces being drilled should be securely fastened to the drill press table to avoid spinning, rolling, or sliding.
2. If a piece of metal does begin to turn on the twist drill, apply pressure on the drill press handle and shut off the power.
3. All metal chips should be removed carefully to avoid injury.
4. Gloves should be worn, but loose clothing, jewelry, and watches should not be worn. Long hair should be kept under a net or cap.

Drilling, Tapping, and Threading: A Review

In drilling holes with a drill press, it is especially important to properly lay out and measure the metal being worked on so that holes to be drilled are correctly located. It is also important to make sure that all safety precautions are taken while the drill press is being used.

When tapping and threading are done, it is vital to select and use the correct tools so that the resulting threads will be correct. Remember to remove all cuttings resulting from tapping and threading.

THINKING IT THROUGH

1. List and identify the basic tools used in metal layout.
2. Describe how you would determine the exact center of a rectangular piece of metal and a piece of round stock.

3. Why is center punching prior to drilling important?
4. Why should metal be clamped securely prior to drilling?
5. Why is the twist-drill speed important?
6. List the common types of threads and describe a typical application for each.
7. How do bolt threads differ from pipe threads?
8. Describe the procedure for threading a rod.
9. Describe the procedure for threading pipe.
10. Describe the procedure for tapping a hole.
11. How do you remove a broken bolt?

CHAPTER 13

FORMING AND FASTENING COLD METAL

Imagine being in this situation: You want to cut a small steel rod, but you don't have access to a saw at the moment. What do you do?

Before this question is answered, imagine another situation: You want to join two pieces of metal, but the job calls for a more permanent and tighter fit than you can get with a bolt or metal screw. What do you do?

The answer to the first question is that you use a cold chisel. The answer to the second question is that you can rivet the metal together instead of bolting it.

Chapter 11 talked about cutting and shaping cold metal by using saws and other fairly sophisticated equipment. And Chapter 12 described joining metal with bolts and screws. This chapter will talk about other and simpler methods for cutting and forming metal and how to fasten metal with rivets instead of bolts and screws.

When five different groups of agricultural workers were asked whether they needed to know the skills of using a cold chisel, bending cold metal, and riveting, most said they needed to know all these skills (see Table 13-1).

CHAPTER GOALS

In this chapter your goals are:

- To perform the necessary layout prior to forming or fastening metal
- To demonstrate how to cut and bend flat stock, round stock, and pipe
- To be able to use a metal bender and a hydraulic pipe bender
- To rivet with both cold and pop rivets
- To identify common bolts, washers, and nuts for specific jobs
- To demonstrate how to do these processes safely

Layout

Layout for chisel cutting and bending of cold metal usually is not quite as detailed as it might be for more sophisticated cutting. In fact, it is hard to be precise in bending. However, errors can be reduced if a few basic procedures are followed.

First, it is necessary to understand the overall project plan and how the different pieces being worked on fit the plan. It may

TABLE 13-1.

	FARMER AND RANCHER	SUPPLIES WORKER	TRACTOR MECHANIC	PACKING HOUSE WORKER	GREEN-HOUSE WORKER	LOGGER
Demonstrate safety while working	X	X	X	X	X	X
Layout on cold metal	X	X	X			
Cutting with a chisel	X	X	X	X		X
Forming cold metal	X	X	X			
Bending pipe	X	X	X	X		X
Riveting	X	X	X	X		X
Selecting fasteners	X	X	X	X		X

be desirable to use a piece of heavy wire or light-gauge metal as patterns to see how the pieces can be formed. In working on sheet metal, the pattern for making different shapes can be made of paper.

When a piece of metal is bent, the outside edge of the metal stretches, and the metal on the inside of the bend compresses somewhat. How well the metal will bend—stretch and compress—depends on the kind of metal being used. A piece of copper tubing, for example, is more ductile than a piece of steel. Naturally, small, thin pieces of metal are more ductile than large, thick pieces. If iron has been alloyed with other metals, the ductility of those metals will have been changed.

A center punch, scribe, or soapstone

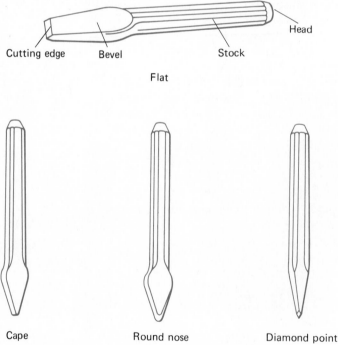

Flat

Cutting edge   Bevel   Stock   Head

Cape   Round nose   Diamond point

Figure 13-1.   Four types of cold chisels.

METAL WORKING

can be used to make reference measurements and marks for cutting with a cold chisel or for bending metal. Measurements and marks for bending should be as accurate as possible.

Cutting Cold Metal

Many light cutting jobs, such as the cutting of mild steel, can be performed well with a cold chisel or bolt cutter.

Types of Cold Chisels

There are four types of cold chisels: *flat*, *diamond point*, *cape*, and *round nose* (see Figure 13-1). The flat chisel is designed for cutting sheet metal, round stock, bolts, nuts, and rivets. The diamond-point chisel is used for cutting square grooves. The cape chisel is designed for cutting rectangular grooves or channels. The round-nose chisel is used to cut grooves with an oval shape.

Using the Cold Chisels

To cut metal that is $1/8$ in [3.2 mm] thick or less, the metal should be secured in a vise. The cutting line that was measured and

marked should protrude slightly above the jaws (see Figure 13-2). The flat chisel is held at a 60° angle to start shearing at one end with a firm strike from a machinist's hammer.

Metal thicker than $1/8$ in [3.2 mm] should be cut by using the chipping block of an anvil (to avoid damage to the hardened face of the anvil). Cuts are begun at both edges, and work progresses toward the center. The metal is then turned over, and the cutting into the edges continues on the other side. After a deep notch has been cut starting from the edges, the metal can be broken by bending it back and forth. Then the rough edge should be ground or filed to eliminate sharp burrs.

Round bar stock can be cut by using the same basic techniques. With a flat chisel, a notch is cut approximately one-third of the way through the stock. The stock is then turned over, and a notch is cut on the opposite side. The round stock should break if it is bent in a vise. Another technique is to use an *anvil hardy*, which is a heavy-duty cutting surface fitted into the square hole in the face of the anvil. The round stock is positioned on the hardy (see Figure 13-3) and struck with a 32- or 40-oz [0.9- or 1.1-kg] machinist's or blacksmith's hammer. The striking is continued until a

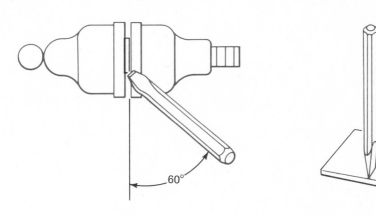

Cutting angle

Starting cut

Figure 13-2. Cutting with a cold chisel.

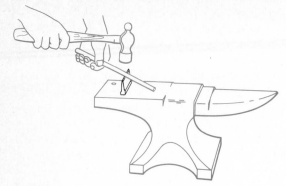

Figure 13-3. Cutting round stock using an anvil hardy.

notch is formed approximately one-third of the way through the stock. The rod is turned over, and the process is repeated. Then the rod can be placed in a vise and bent until it breaks clean.

Using Bolt Cutters

Bolt cutters can be used to cut round metal stock up to ¾ in [19.1 mm] in diameter (see Figure 13-4). Usually the manufacturer will indicate both the maximum capacity and the maximum hardness of stock for which the cutter is designed.

The metal is placed as far to the rear of the jaws as possible, and the mark must be made on the cutting edge. Then the handles are pulled together without allowing the cutter to twist the stock. If the jaws start to twist, they can be damaged. When large-capacity cutters are used, one handle can be placed on the floor while the other is forced down.

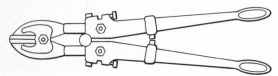

Figure 13-4. Bolt cutters are often used to cut round stock.

Using Other Hand Tools

Many light-gauge mild-steel and wrought-iron pieces can be bent or shaped with a vise, anvil, and such hand tools as the blacksmith's and machinist's hammers.

After examination of the sketch or plan, the necessary measurements and marks are made. If the jaws of the vise are knurled (crosshatched), a smooth liner (soft jaw) should be used to avoid damaging the metal.

Either procedure can be used to make right-angle bends. In one method, a 32-oz [0.9-kg] blacksmith's hammer is used to strike the metal sharply just above the jaws of the vise (see Figure 13-5). The metal is struck until the proper angle is formed. The angle can be checked with a square, T bevel, or template. The other method is to slip a long piece of pipe over the piece to be bent and to do the bending as you would bend hot metal (see Chapter 14).

If it is desirable to form an eye (an

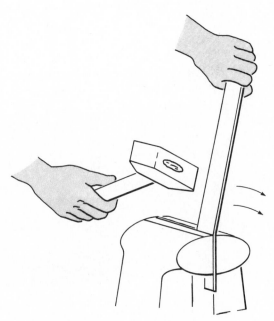

Figure 13-5. Forming right angle bends.

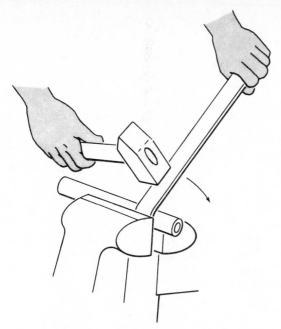

Figure 13-6. Forming the centered eye.

such as artistic wrought-iron stair railings. If it is properly done, twisting the metal actually will increase its strength. Because twisting tends to shorten the piece of metal, it is desirable to cut the piece to exact length after the twisting is completed.

To begin the twisting, the piece of metal is clamped firmly in the jaws of a metal-working vise. If necessary, liners (soft jaws) should be used to protect the finish of the metal. An adjustable open-end wrench is used to twist the metal the desired number of times. If a long piece of metal is being twisted (as for a stair railing), a close-fitting pipe can be placed over the piece to prevent buckling.

Metal Bender

Many shops are equipped with a metal bender. This is a very handy tool, and with it an experienced worker can create a variety of metal shapes. Only some basic shapes such as eyes, circles, and zero-radius angles will be discussed here.

Most benders will accept flat, round, channel, and angle iron. Usually, mild steel and wrought iron are used. Other kinds of steel and cast iron are too hard and tend to break if bent. Even cold-rolled steel does not usually produce a desirable bend because it is not very ductile.

To form an eye with the bender, first the diameter of the eye must be determined. A radius collar of the same diameter is then selected. While the metal is held against the collar with a locking pin, the arm of the bender is turned clockwise against the stock to begin the eye (see Figure 13-7). This process is continued until the end of the metal closes the eye. To make a centered eye, the sweep of the arm is continued, forcing the leg of the eye bolt to intersect the center line of the eye.

enclosed circle) at the end of a piece of flat stock, a piece of round stock is selected that has the same diameter as the eye desired. Both the flat and the round stock are placed in a vise, as shown in Figure 13-6. About half the round stock should show above the jaws.

With one hand, pull the metal over the round stock. Hammer the back of the metal in the same direction. This will cause it to form the desired radius. Loosen the vise and lower the metal in the jaws (still against the round stock). Tighten the vise and continue to form the metal to the radius of the round stock.

Finally, place the U-shaped eye in the vise and strike the end of the metal until the eye is closed. The eye can be centered by striking the metal from the opposite side and bending it back over the jaw of the vise.

Sometimes it is desirable to twist metal for special needs or special effects,

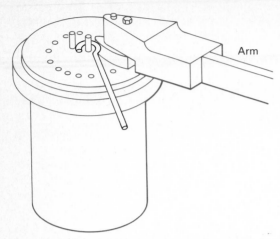

Figure 13-7. Forming an eye from round stock with a metal bender.

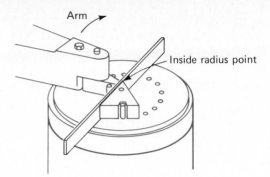

Figure 13-9. Forming a zero-radius bend with a metal bender.

A circle is formed by first determining the diameter and then figuring the length of the metal needed. The length is determined by multiplying 3.14 times the diameter of the circle. Since there is resistance to bending in most metals, a radius should be selected that is slightly smaller than the desired circle. The center of the metal is clamped against the radius collar, and the arm is advanced until it reaches the end of the stock (see Figure 13-8). The material is then relocated by turning the stock over and the bending operation is repeated. If the circle is to be welded into one piece, it may be necessary to clamp it lightly into a vise prior to tack welding.

If a sharp angle or a zero-radius bend is desired in a piece of metal, the ductility of the metal must be considered. Usually, only low-carbon steel and wrought iron are used. Most benders have a zero-radius block that is used to form a sharp angle. The desired angle and the location of the bend are determined first. Then the metal is positioned in the bender so that the inside radius fits against the block (see Figure 13-9). The arm is advanced until the desired angle is reached. Before the metal is removed, the angle should be checked with a square, T bevel, or template. If the angle is correct and it is desirable to duplicate the piece, a mark can be made on the base of the metal bender to serve as a future guide.

Pipe Bender

The worker on a livestock farm may want to construct a hog farrowing crate, where pigs are born, using pipe. Some of the pipe will have to be bent. That's when one should select the pipe bender. The pipe bender is available as either a mechanical power unit or a hydraulic power unit. Because of the pressure required to form steel pipe, most benders are hydraulic.

The bender unit consists of a hydrau-

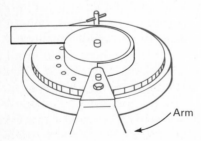

Figure 13-8. Forming a circle from flat stock with a metal bender.

METAL WORKING

lic pump, which may be manual or powered by a motor; a hydraulic cylinder; and a frame. There are two types of hydraulic benders: the *segmental* (requiring several sets to complete a bend) and a *one-shot*, or *sweep, bender*. The sweep bender is the most popular for forming pipe under 3 in [76.2 mm] in diameter.

In working with high-pressure hydraulic equipment, there are some safety practices to follow carefully. To begin with, the load limits of the pump or cylinder should not be exceeded. Also, the travel limits of the cylinder plunger should not be surpassed. These limits are identified by the manufacturer. Sharp bends or kinks in the hose should be avoided, and the operator always should stand behind the bender frame when forming pipe. A great deal of force is required to make the bend, and the operator could be seriously hurt if the pipe came out of the frame.

Pipe also tends to stretch on the outside of the bend and compress on the inner radius. Because of this, the manufacturer makes specific recommendations regarding the preparations for making a 90° angle. The operator's manual should be checked for the proper setback dimension before work is begun on a machine.

To operate the sweep bender, first the size of the pipe and the angle to be formed are determined. Then the proper size on the swivel shoe and bending shoe is selected. The pipe is inserted into the bender, and the layout mark is aligned with the center line of the bending shoe. The pump is operated until the pipe is seated between the swivel and bending shoes. The angle gauge reading of 00° must align with the reference mark on the frame. If this is correct, the pumping is continued until the angle gauge registers the desired angle. Since pipe tends to spring back after the pressure is released, it may be necessary to *slightly* overbend the piece. The

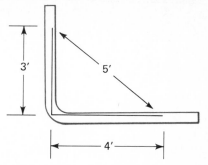

Figure 13-10.   Checking a 90° bend with a 3-4-5 triangle.

pressure is released slowly, and the bend angle should be checked. If the angle is satisfactory, the pipe can be removed from the bender frame.

If the angle is overbent, most manufacturers have a correcting procedure. Generally the pipe is reversed in the frame and the swivel shoes are relocated. This will apply pressure on the outside radius of the bend, and that should correct the problem. The overbend must be reduced gradually along the full length of the bend. Otherwise, the pipe may break.

After the pipe has been removed from the frame, it is a good idea to recheck the angle with a framing square or a 3-4-5 triangle (see Figure 13-10).

Fastening Cold Metal with Rivets

Rivets are more permanent than bolts and screws. They also are used when the head of a bolt or screw might not allow enough clearance.

Rivets used in agricultural industry are grouped into two general classes: *cold rivets* and *pop rivets*. Both are available in a variety of head types. The most common types are button, flat, tinner's, pan, and countersunk head (see Figure 13-11).

In using rivets, it is important to know

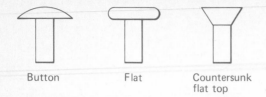

| Button | Flat | Countersunk flat top |

Figure 13-11. Three common rivet heads.

the correct hole diameter. The rivet must fit very tightly. If it doesn't fit snugly, eventually the metal will loosen and shear the rivet.

Using the Cold Rivet

To join two pieces with *cold rivets*, first the centers of the rivet holes are carefully marked and center punched. The hole is drilled with the same size twist drill as the rivet. The rivet selected should extend from $1/16$ to $1/8$ in [1.6 to 3.2 mm] above the surface of the metal. This will provide adequate material with which to form the head. Be sure that the two metal pieces fit tightly before you start to set the head. A rivet set is a handy tool to ensure a proper fit.

The first few hammer blows should be directed straight down on the rivet, which causes the shank to expand. The final blows should be made with the peen, or rounded, end of the hammer to form the head. The head should not be flattened, because most of the rivet's strength comes from the round head. Some rivet sets have a cup-shaped hole which helps form the rivet head (see Figure 13-12).

Using the Pop Rivet

The *pop rivet* is a very versatile fastener that is fast and easy to use. Its most interesting feature is its ability to secure two or more pieces of metal together when it is possible to work from one side only. In other words, it can be used in blind locations.

Although it does not resist pulling apart or shearing as well as either a cold rivet or a bolt, the pop rivet performs very well. The pop rivet is sized by the diameter and the grip length of the head. Pop rivets are available commonly in diameters of $3/32$, $1/8$, $5/32$, $3/16$, and $1/4$ in [2.4, 3.2, 4.0, 4.8, and 6.4 mm]. Grip lengths range from

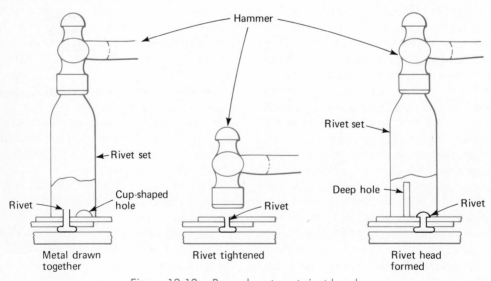

Figure 13-12. Procedure to set rivet heads.

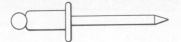

Figure 13-13. Pop rivets are used to fasten many sheet metal projects.

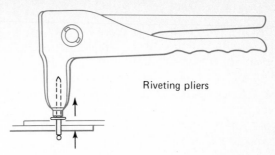

Riveting pliers

Figure 13-14. Using pop riveting pliers.

$^5/_{32}$ to $^{15}/_{16}$ in [4.0 to 23.8 mm]. The protruding head of the pop rivet should be equal to 20 to 25 percent of the rivet's length. Each pop rivet will have a grip range specified by the manufacturer. Pop rivets also are available with several types of heads. The most common is the *button head* (see Figure 13-13). This is used when it is desirable to have a flush or smooth surface. *Countersunk* and *large-flange heads* also are available for special applications.

It is important to select the correct type of rivet for the metal being used.

Some metals will react with other metals to cause corrosion and rivet failure. Different rivets are available for fastening steel, aluminum, copper, and stainless steel. The instructor or supplier will help in the selection.

Basically, the same considerations apply in using either pop or cold rivets. The location of the hole should be identified with a center punch and then drilled to the correct size. Incorrect drill size accounts for more pop rivet failure than all other factors combined. For example, one manufacturer specifies no more than 0.004 in [0.1 mm] of tolerance on the diameter of the hole. It is possible to have a greater error than that by using an improperly conditioned twist drill.

To install a pop rivet, the pin of the rivet is inserted into the nose piece of riveting pliers. Then, the body of the rivet is pressed firmly into the hole (see Figure 13-14). Make sure that the two pieces of metal fit tight. Then lock the rivet in place by closing the pliers with a long, smooth stroke. Occasionally, several strokes may be needed.

Bolts, Screws, and Nuts

Bolts, screws, and nuts also come in a variety of shapes and sizes. The thread characteristics of bolts and screws are discussed in Chapter 12. Fasteners may include self-tapping screws, machine screws, and cap screws, as well as common bolts.

Self-tapping screws are sized according to the American wire gauge (AWG) and are numbered 2, 4, 6, 7, 8, 10, 12, and 14. The largest number is the largest in diameter. As the name implies, a self-tapping screw does not require a threaded hole. It makes its own thread as it feeds into the hole. Because it has coarse threads with a shallow throat, it is not designed for uses where there will be frequent assembly and disassembly. The common head types include flat, round, oval, pan, and truss (see Figure 13-15).

Machine screws are available in numbered sizes ranging from 2 through 12 and as fractional diameters from $^1/_4$ to $^3/_8$ in [6.4 to 9.5 mm]. Machine screws are also available both in metric sizes and in NF (national fine) and NC (national coarse)

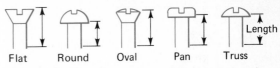

Flat Round Oval Pan Truss

Length

Figure 13-15. Common machine screw heads.

thread standards. For example, a screw designated by the numbers 8–32 would be a number 8 wire gauge diameter with 32 threads per linear inch. The machine screw is used to join machine parts and light-gauge metal where disassembly may be necessary later. Of course, they require a threaded hole or a nut for fastening.

The *cap screw* is essentially a threaded bolt that is used in a threaded hole. It has a hexagonal head and is commonly available in sizes ranging from ¼ to 1 in [6.4 to 25.4 mm]. Cap screws are available in a variety of tensile strengths. A cap screw may have an NC, NF, ISO coarse, or ISO fine thread.

Common bolts are used widely as fasteners. The bolts come in many different sizes and head types (see Figure 13-16). There is a bolt designed for every fastening job. Common sizes range from ¼ to 1 in [6.4 to 25.4 mm] in diameter and from ½ to 20 in [12.7 to 508 mm] in length. Bolts also are available with a variety of head types. The most common are *hexagon*, *square*, *carriage*, and *plow bolts*.

Washers and Nuts

Washers are important to fastening metal pieces (see Figure 13-17). The two common washers used in agricultural industry

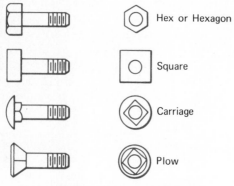

Figure 13-16. Common bolt heads.

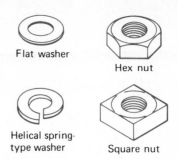

Figure 13-17. Typical nuts and washers used to fasten metal pieces.

are the *flat* and *lock washers*. The flat washer is used to provide a larger bearing surface for the head of the bolt or nut. Therefore, the washer can be used to prevent the head of the bolt or screw from pulling through the metal. The lock washer is designed to provide continuous pressure on the bolt assembly. As the nut tries to loosen, the edge of the lock washer digs into the base of the nut.

All bolts require a nut to complete the fastening operation. Nuts come in many different types to complement the variety of bolts. The most common nuts in agricultural industry are the hexagonal and, to a lesser extent, the square nuts. The hexagonal (hex) nuts are further subdivided into standard hexagonal, hexagonal jam, hexagonal castle, and the hexagonal slotted nuts.

The width of the standard hex nut from one flat side to another is usually 1.5 times the diameter of the hole. If very large holes are used, a heavy hex series is available with a wider flat.

The hex jam nut is approximately one-third the thickness of the standard nut, and it is used only for locking and for light-duty applications. The castle and slotted nuts permit the use of a cotter pin or a safety wire for locking into position. These are used mostly on power-drive lines and with internal machine and equipment parts. Nylon and other synthetic materials are available to insert onto the threads for special jobs.

The square nut was common on machinery with low torque requirements, but it has largely been replaced by the hex nut.

Safety Rules

These safety rules should be followed in forming and fastening cold metal:

1. Always wear safety goggles and gloves.
2. An air hose should not be used to blow away metal particles from the work area.
3. Keep the work area free of extra tools and scrap metals.

Forming and Fastening Cold Metal: A Review

Metal can be cut without being sawed. One way to cut cold metal is with a cold chisel, but it is important to select the right chisel for the specific job.

It is often necessary to bend or shape cold metal. Sometimes this can be done in a vise, but on other occasions a special tool such as a metal bender or pipe bender must be used. Very high pressures are used in a hydraulic bender, and the operator must guard against injury.

Inserting cold or pop rivets is another way of fastening metal. Rivets are more permanent fasteners than either bolts or screws.

THINKING IT THROUGH

1. List the tools needed to lay out metal for cutting or bending.
2. Identify the four kinds of cold chisels and describe how each could be used.
3. Describe how metal to be cut with a chisel should be fastened in the metal-working vise.
4. Explain how metal can be protected when it is fastened in the jaws of a metal-working vise.
5. List the safety precautions necessary in reshaping and fastening cold metal.
6. List several examples of where pop rivets would be the ideal way to fasten metal.

CHAPTER 14

WORKING HOT METAL

While some metals can be worked when they are cold, others can be worked best when they are hot. Most agricultural workers who form and shape cold metal also have to heat metal, either to reshape it or to heat-treat it. The job may be heating a heavy part that needs to be straightened or taking a good tool that has been previously tempered (hardened) and needs to be heat-treated and retempered.

CHAPTER GOALS

In this chapter your goals are:

- To demonstrate how to heat metal in a gas furnace until it is at the correct temperature for reshaping or heat treating
- To identify tools to use in these processes
- To demonstrate how to heat metal and temper and retemper tools and equipment parts
- To be able to work safely with hot metal

Heating Metal for Reshaping

Metal that has to be reshaped is heated either in a furnace or with an oxyacetylene heating torch. Heating metal in a gas furnace will be discussed here. Heating metal with a torch is discussed in Chapter 22.

Using the Gas Furnace

The gas furnace is available as a pedestal (or bench) model and as a floor model (see Figure 14-1). The floor model has a greater capacity than the pedestal model and can heat a number of pieces at one time. However, it will not take long pieces. The pedestal furnace, on the other hand, has a very limited capacity. Because it is open at both ends, it can handle a long piece that is fed into it for heating in specific locations on the piece.

Both models heat to approximately 3000°F [1649°C]. The gas furnace has an important advantage over both the coal forge and the oxyacetylene torch because it can heat metal for long periods without actually burning or oxidizing the metal.

When either furnace is used, it is important to follow the lighting instructions provided by the manufacturer. The gas furnace should be equipped with an electrical solenoid valve that prevents gas from passing into the fire box unless the blower is operating. This avoids the possibility of gas accumulating in the fire box and exploding.

Heating Ferrous Metals

Much can be learned about heating different metals at different temperatures by studying Table 14-1. Almost all iron and steel must be heated to 1200°F [649°C] or above before they take on the dark-red

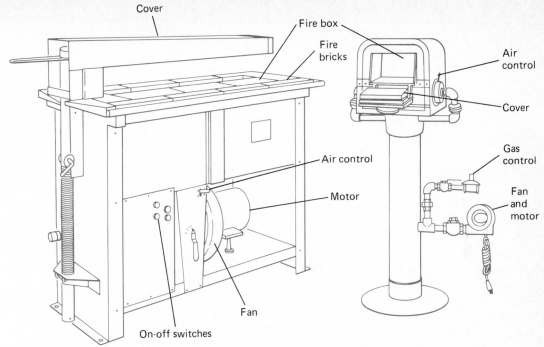

Figure 14-1. The gas furnace is available in either a floor or pedestal model.

color that indicates that the metal is ready for reshaping. The grain structure of metal remains the same up to about 1300°F [704°C]. Above this level the grain size becomes smaller until a temperature of 1800°F [982°C] is reached. Then the grain size enlarges again. It continues to enlarge until the metal melts, usually at about 2800°F [1538°C]. Grain size determines how hard or brittle the metal will be if it is cooled rapidly after being heated to a temperature above 1300°F [704°C].

Another thing to note is that steel, when heated, is magnetic up to 1400°F [760°C]. Above this temperature, steel loses its magnetic quality.

Tools Used in Working Hot Metal

Tools used to work hot metal are the anvil, vise, blacksmith tongs, hammer, hardy, punch, and steel measuring rule.

Anvil

The *anvil* (see Figure 14-2) is a very versatile piece of shop equipment. The face and horn of the anvil are hardened and used for the shaping of hot metal. The chipping block does not have a hardened surface, and it can be used for working cold metal with a cold chisel. The anvil should

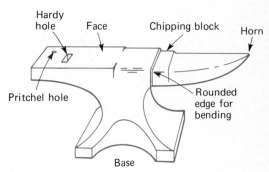

Figure 14-2. The blacksmith's anvil is available in a variety of weights.

TABLE 14-1. Color-Temperature Chart of 0 to 90 Point Carbon Steel

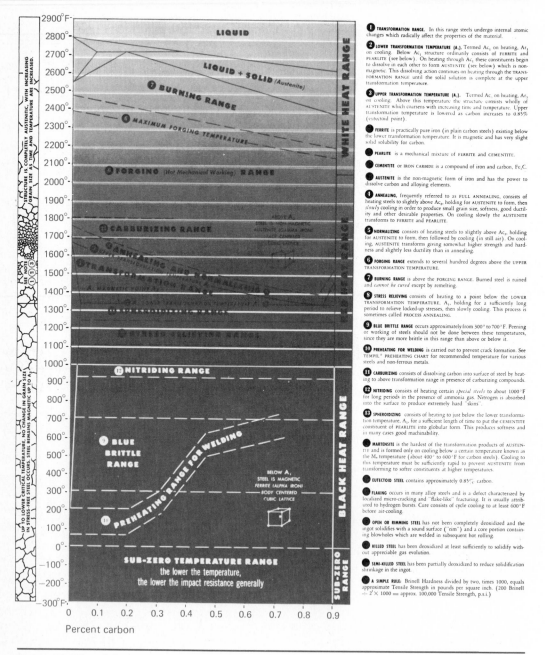

1 TRANSFORMATION RANGE. In this range steels undergo internal atomic changes which radically affect the properties of the material.

2 LOWER TRANSFORMATION TEMPERATURE (A_1). Termed Ac_1 on heating, Ar_1 on cooling. Below Ac_1 structure ordinarily consists of FERRITE and PEARLITE (see below). On heating through Ac_1 these constituents begin to dissolve in each other to form AUSTENITE (see below) which is non-magnetic. This dissolving action continues on heating through the TRANSFORMATION RANGE until the solid solution is complete at the upper transformation temperature.

3 UPPER TRANSFORMATION TEMPERATURE (A_3). Termed Ac_3 on heating, Ar_3 on cooling. Above this temperature the structure consists wholly of AUSTENITE which coarsens with increasing time and temperature. Upper transformation temperature is lowered as carbon increases to 0.85% (eutectoid point).

FERRITE is practically pure iron (in plain carbon steels) existing below the lower transformation temperature. It is magnetic and has very slight solid solubility for carbon.

PEARLITE is a mechanical mixture of FERRITE and CEMENTITE.

CEMENTITE or IRON CARBIDE is a compound of iron and carbon, Fe_3C.

AUSTENITE is the non-magnetic form of iron and has the power to dissolve carbon and alloying elements.

4 ANNEALING, frequently referred to as FULL ANNEALING, consists of heating steels to slightly above Ac_3, holding for AUSTENITE to form, then slowly cooling in order to produce small grain size, softness, good ductility and other desirable properties. On cooling slowly the AUSTENITE transforms to FERRITE and PEARLITE.

5 NORMALIZING consists of heating steels to slightly above Ac_3, holding for AUSTENITE to form, then followed by cooling (in still air). On cooling, AUSTENITE transforms giving somewhat higher strength and hardness and slightly less ductility than in annealing.

6 FORGING RANGE extends to several hundred degrees above the UPPER TRANSFORMATION TEMPERATURE.

7 BURNING RANGE is above the FORGING RANGE. Burned steel is ruined and *cannot be cured* except by remelting.

8 STRESS RELIEVING consists of heating to a point below the LOWER TRANSFORMATION TEMPERATURE, A_1, holding for a sufficiently long period to relieve locked-up stresses, then slowly cooling. This process is sometimes called PROCESS ANNEALING.

9 BLUE BRITTLE RANGE occurs approximately from 300° to 700°F. Peening or working of steels should not be done between these temperatures, since they are more brittle in this range than above or below it.

10 PREHEATING FOR WELDING is carried out to prevent crack formation. See TEMPIL® PREHEATING CHART for recommended temperature for various steels and non-ferrous metals.

11 CARBURIZING consists of dissolving carbon into surface of steel by heating to above transformation range in presence of carburizing compounds.

12 NITRIDING consists of heating certain *special steels* to about 1000°F for long periods in the presence of ammonia gas. Nitrogen is absorbed into the surface to produce extremely hard "skins".

13 SPHEROIDIZING consists of heating to just below the lower transformation temperature, A_1, for a sufficient length of time to put the CEMENTITE constituent of PEARLITE into globular form. This produces softness and in many cases good machinability.

MARTENSITE is the hardest of the transformation products of AUSTENITE and is formed only on cooling below a certain temperature known as the M_s temperature (about 400° to 600°F for carbon steels). Cooling to this temperature must be sufficiently rapid to prevent AUSTENITE from transforming to softer constituents at higher temperatures.

EUTECTOID STEEL contains approximately 0.85% carbon.

FLAKING occurs in many alloy steels and is a defect characterized by localized micro-cracking and "flake-like" fracturing. It is usually attributed to hydrogen bursts. Cure consists of cycle cooling to at least 600°F before air-cooling.

OPEN OR RIMMING STEEL has not been completely deoxidized and the ingot solidifies with a sound surface ("rim") and a core portion containing blowholes which are welded in subsequent hot rolling.

KILLED STEEL has been deoxidized at least sufficiently to solidify without appreciable gas evolution.

SEMI-KILLED STEEL has been partially deoxidized to reduce solidification shrinkage in the ingot.

A SIMPLE RULE: Brinell Hardness divided by two, times 1000, equals approximate Tensile Strength in pounds per square inch. (200 Brinell ÷ 2 × 1000 = approx. 100,000 Tensile Strength, p.s.i.)

Source: The Tempil Corp.

METAL WORKING

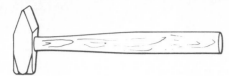

Figure 14-3.   A blacksmith's hammer.

weigh about 150 lb [68 kg] and should be made of steel rather than cast iron. When struck lightly with a hammer, a steel anvil has a distinct ring to it. The cast-iron anvil has a muffled sound when struck with a hammer. The face of the anvil should never be struck with a hammer since both are hardened and will fracture.

Hammer

The blacksmith hammer (see Figure 14-3) comes in several sizes. The typical shop needs one weighing 2 lb [0.9 kg] and another weighing 2½ lb [1.1 kg].

Hardy

The *hardy* is shaped like a chisel (see Figure 14-4), and it fits into a square hole on the face of the anvil. The hardy can be used for cutting either hot or cold metal. In cutting on the hardy, the metal is laid over it and struck with a blacksmith's hammer. The metal should be hammered from both sides, but the hammering should stop before the metal is cut through. This prevents the hardy blade from being damaged.

Punch

The punch shown in Figure 14-5 should be equipped with a handle similar to a ham-

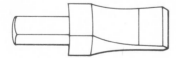

Figure 14-4.   The hardy is used to cut hot or cold metal on the anvil.

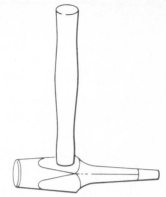

Figure 14-5.   The round eye punch is used to form holes in hot metal.

mer handle. The punch is used to make small holes in hot metal. It is done by positioning the white-hot metal over the pritchel hole on the face of the anvil (see Figure 14-2) and then hammering the punch until it goes through the metal and into the pritchel hole.

Tong

An assortment of tongs (see Figure 14-6) is needed. The different-shaped jaws allow round, flat, and curved pieces of hot metal to be grasped firmly and safely.

Vise

A heavy-duty machinist's *vise*, as seen in Figure 14-7, is needed to hold hot metal for bending.

The steel measuring tape or ruler is used to measure hot metal.

Cutting and Bending Hot Metal

Wrought iron, mild steel, and tool steel can all be heated and then cut, bent, drawn, or upset. The softer metals such as brass, lead, and copper can be melted and reshaped, but usually they are not heated for the purposes previously stated.

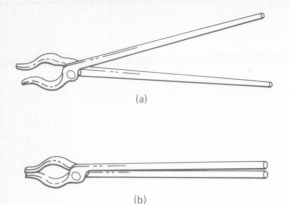

(a)

(b)

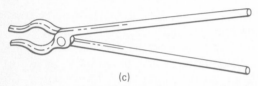

(c)

Figure 14-6. The curved lip (a), flat lip (b), and gad (c) are blacksmith's tongs used for handling hot metal.

In almost all operations, the metal is heated until it reaches a bright-orange color. This usually occurs at 2000°F [1093°C]. The reshaping must begin quickly, before the metal cools.

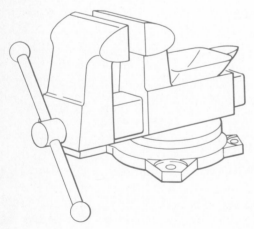

Figure 14-7. A heavy-duty machinist's vise with an anvil.

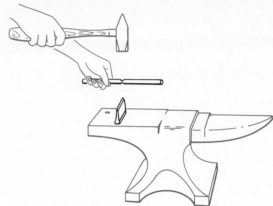

Figure 14-8. Cutting hot metal using a hardy.

To cut metal after it is heated, the metal is placed over the hardy on the anvil. Then the metal is struck on each side to cut it (see Figure 14-8). Instead of cutting all the way through and possibly damaging the hardy, go only far enough that the metal can later be bent and separated.

Bending usually refers to the reshaping of metal into square and angle bends and other positions. Square and angle bends are made by heating the metal, placing it on the face of the anvil, and striking it with a hammer until the desired angle is achieved (see Figure 14-9). A steel square or metal T bevel can be used to check the angle.

Hot metal also can be bent over the horn of the anvil. One use for the horn is to

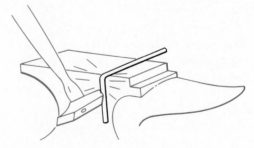

Figure 14-9. The face of the anvil should be used to form a right-angle bend on hot metal.

METAL WORKING

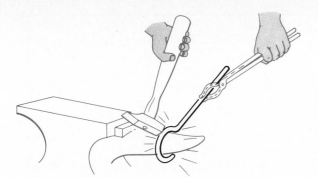

Figure 14-10. The horn of the anvil is used to form an eye or other curved shapes.

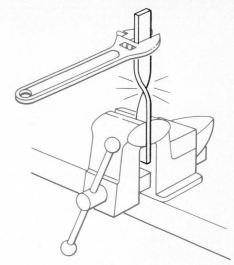

Figure 14-11. Use an adjustable end wrench and a vise to twist hot metal.

bend metal to create an eye (see Figure 14-10). Other curved angles and shapes also are possible.

Round stock that has been properly heated to a bright orange can be bent in a vise. After you place the metal in the vise, slip a pipe over the metal. The pipe is pulled to begin bending the metal. A very sharp and accurate bend can be made in this way.

Hot metal also can be twisted. First, the metal is placed into the vise as one would do for bending. Then, using a large, adjustable wrench, begin twisting the metal (see Figure 14-11).

Upsetting metal means to reduce its length and increase it diameter. It makes the metal thicker and shorter. To upset metal, the piece is heated until it is orange. Then dip, or quench, the portion of the metal that will not be upset into cold water. Place the still-hot portion of the metal on the face of the anvil, and begin hammering on the cooled portion. The force will spread the heated portion.

Heat Treating Carbon Steel

Steel is heated not only for the purposes mentioned, but also for treating in certain ways. For example, an agricultural worker may have to heat metal in order to increase its ductility (ability to be shaped), to make it more easily machined, to relieve stresses in the metal, to change the grain size, to increase the hardness or softness, or to make it tougher.

There are four processes in heat-treating steel: *heating*, *annealing*, *hardening*, and *tempering*. Most steel tools can be treated because they have carbon percentages above 0.50 (50 points). Items such as nails, pipes, and wire contain small amounts of carbon (0.10 percent or 10-point) and cannot properly be hardened through heat treatment. Tools such as cold chisels, set screws, and wood saw blades contain 0.75 to 0.85 percent carbon (75 to 85 points). Other tools such as files, lathe tools, metal saw blades, and springs contain 1 to 1.50 percent carbon (100 to 150 points). The maximum amount of carbon that can be combined with steel is about 1.75 percent (175 points). Of course, the higher the percentage of carbon, the harder, the stronger, and the more brittle the steel is.

A cold chisel can be used to illustrate the various stages of heat treatment. A properly made cold chisel has a hardened, or tempered, cutting edge. The part directly behind the cutting edge is softer than the cutting edge, but still hard. The shank, or hammer, end (which is struck by a hammer) is softer and more ductile. This prevents the end from shattering when struck with a hammer.

The chisel is first heated and shaped. Once the chisel has been formed, the entire tool should be annealed in preparation for the other steps. To *anneal* means to heat the metal until it is relieved of stress. In the case of the chisel, it should be heated to a cherry-red color (about 1350°F, or 732°C). Then, the chisel should be placed immediately into a bucket of dry sand, ashes, or lime. It should remain in this material until it has cooled below 1000°F [538°C] or until all color has disappeared. Metal can be left in the material until completely cold, if preferred.

After the annealing process, the chisel is again reheated for about two-thirds of its length (starting from the cutting edge). Reheat to above 1350°F [732°C] or until the steel is again cherry red (refer to Table 14-1). While it is still cherry red, quench —dip—approximately 1½ in [3.8 cm] of the chisel point into water until it loses all of its color. Withdraw the chisel from the water while the remainder of the tool is still cherry red.

Quickly polish the chisel point with an emery cloth or a mill file until it is shiny. One must be careful because the point is hot. Now the heat in the remainder of the chisel will begin to spread down into the point. As this happens, the shiny point will turn different colors (see Figure 14-12).

The first color to appear is yellow. This indicates a temperature of about 430°F [221°C]. Quenching the point at this stage would produce an extra-hard chisel. Following the yellow color, in order, will be straw, brown, purple, and blue. The blue color means that the point has reached a temperature of about 560°F [293°C]. When the cutting edge turns straw, it is the best time for quenching. This will give the edge a proper temper—not so hard that it will be brittle and easily fractured.

To quench the cutting edge, dip it repeatedly in water. During this process, the remainder of the chisel should not be quenched. Otherwise, the shank could become too hard. Only when the red color has left the chisel can the entire chisel be cooled.

The correct tempering colors vary among tools. Punches, knives, lathe tools, and drills, for example, should be tempered when their points turn a straw color. A brown color is best for axes and wood chisels. Blue is appropriate for screwdrivers.

There are other ways to determine when metal has reached a certain temperature. Special temperature sensing crayons, pellets, and liquids are available that melt

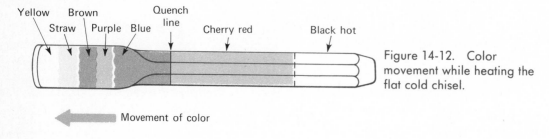

Figure 14-12. Color movement while heating the flat cold chisel.

METAL WORKING

at specific temperatures when exposed to hot metal. These materials can designate temperatures ranging from 113°F [45°C] to 2500°F [1371°C], in steps of 12.5°F [−10.8°C] to 100°F [37.7°C].

In some cases it may be desirable to blacken forged parts to improve their appearance and give some protection from rust. Blackening is done by heating the part just enough so that it will smoke when rubbed with an oily rag.

Safety Rules

These safety rules should be followed when you are working with hot metal:

1. Wear safety goggles, long-sleeved garments, and leather gloves.
2. Make sure the handle of the blacksmith's hammer fits tightly so that the head won't fly off.
3. Handle hot metal only with tongs.
4. Follow the operator's manual for lighting the furnace, and make sure no flammable materials are near the furnace.
5. Keep the work area free of excess tools and scrap materials.

Working Hot Metal: A Review

Workers in most agricultural career fields need to know how to work with both cold and hot metal.

For example, certain metals are difficult to reshape unless they are heated. And heat is necessary to treat steel tools such as chisels.

It is important to learn the different temperatures at which different metals may be shaped, tempered (hardened), and annealed. It is also important to know how to properly cool heat-treated metal.

THINKING IT THROUGH

1. Prepare a list of items from your home or farm that need to be made or repaired through heat treatment.
2. List the steps in the tempering of a cold chisel.
3. Describe the effect heating metal has on the grain structure of metal.
4. List the approximate temperatures of steel that is being heat-treated when the steel is in the following conditions: (a) nonmagnetic, (b) dark red, (c) dark orange, (d) yellow, (e) liquid or molten, (f) black/orange.
5. Describe the effect of annealing on a piece of steel.
6. What personal safety equipment is necessary when working with hot metal?

UNIT V

BASIC ARC WELDING

COMPETENCIES

Competency	Production Agriculture (Diversified farmer)	Agricultural Supplies/Services (Fertilizer applicator)	Agricultural Mechanics (Field service mechanic)	Agricultural Products, Processing, and Marketing (Mill maintenance person)	Horticulture (Garden center mechanic)	Forestry (Sawmill operator)	Renewable Natural Resources (Soil conservationist)
Demonstrate personal safety while arc welding	Very Important	Very Important	Very Important	Very Important	Important	Very Important	Important
Identify welding equipment and electrodes	Very Important	Very Important	Very Important	Very Important	Important	Very Important	Important
Weld in the flat position	Very Important	Very Important	Very Important	Very Important	Very Important	Very Important	Important
Weld in the horizontal position	Very Important	Important	Very Important	Important	Important	Important	Not Important
Weld in the vertical position	Very Important	Important	Very Important	Important	Not Important	Important	Not Important
Weld in the overhead position	Important	Not Important	Very Important	Important	Not Important	Not Important	Not Important
Weld a pipe joint	Very Important	Not Important	Very Important	Very Important	Important	Important	Not Important

 Very Important   Important   Not Important

BASIC ARC WELDING

Pieces of metal can be joined by bolts, screws, and rivets, or they can be joined by heat. When two pieces of metal are heated so that the molten metal from each piece flows together, the process is called *fusion welding*. Two pieces also can be joined through a nonfusion welding process called brazing or soldering. In this process, the metals to be joined are heated and a metal alloy filler rod flows between the two pieces, adhering to the surface of the metals to be joined.

The majority of welding in agricultural industry is arc welding. In arc welding, electricity is used to create the necessary heat for fusion welding.

Workers in many agricultural career fields use arc welding. In production agriculture, farm and ranch workers, for example, use arc welding to repair equipment. Much of the welding is necessary to repair a machine which breaks during operation. It may be done in a shop or in the field with a portable arc welder.

Construction companies in agricultural industry use portable arc welding equipment to repair equipment and erect grain-handling structures, metal buildings, and feed mills. The mechanic may use the arc welder to build or repair a piece of equipment, and equipment dealers often repair machinery by using arc welding. Logging and land management workers use engine-driven welders mounted on a service truck to make quick repairs on bulldozers, skidders, and other equipment.

When you have completed this unit, you should be able to identify and use different types of arc welders. You also should be able to prepare metal for welding, perform common welds, and point out problems that might weaken a weld. You also should demonstrate the rules for safe operation of arc welding equipment.

CHAPTER 15

BASIC PRINCIPLES OF ARC WELDING

Metal is frequently used in agricultural industry because it is strong and serves many purposes. Metals such as steel, aluminum, cast iron, stainless steel, and copper are most often used. Steel is commonly used in tools, equipment, and general construction.

Particularly in construction, it is often necessary to join pieces of metal by heating and welding. One way to weld metal is by heating and melting it using electricity. This is called *arc welding*

Before learning how to weld, it is necessary to know some of the basic principles of electricity. A worker also must know the different types of arc welding equipment and be able to operate the equipment safely.

The worker who does not become thoroughly familiar with arc welding equipment may be injured or injure someone else.

CHAPTER GOALS

In this chapter your goals are:

- To describe the arc welding process
- To define the basic electrical terms used in arc welding
- To list and describe the techniques that reduce distortion in metal
- To identify the requirements for a sound weld
- To describe how to evaluate the weld
- To sketch the basic weld joints used in agricultural industry
- To identify the four welding positions
- To apply basic safety practices when welding

Arc Welding is a Fusion Process

Arc welding is a fusion process. That is, metal pieces are joined by heating and melting. The metals are heated until they flow as a molten liquid. The heat is released by an electric current (the arc) flowing between a flux-coated steel rod, called an electrode, and the base metal. The welder is able to control the amount of heat and the flow of molten metal.

The arc welding process begins with the worker striking an arc between the electrode and the base metal. The molten metal from the tip of the electrode forms a bead on the base metal. When two pieces of metal are being welded together, the molten metal forms a joint between the pieces (see Figure 15-1). The molten metal

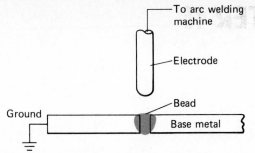

Figure 15-1.   Fusion welding using the arc welder.

from the electrode combines with the molten base metals and hardens to form a single piece.

Basic Electrical Terms

A welder should know the basic principles of electricity. Most arc welding equipment operates on a 240-volt (V) source. The *voltage* is the measure of the electrical "pressure." This force is similar to the pressure which forces water to flow through a pipe. The farm-type welder changes, or transforms, the 240-V source pressure to a much lower electrical pressure at the electrode; usually between 15 and 25 V.

Amperage is the measure of the electrical current flowing through a circuit. In the arc welder it is used as an indication of the amount of heat being produced. Most of the farm-type arc welding equipment is *constant current*. This means that a constant amount of current is flowing between the electrode and the base metal during the welding process. The amount of current available is determined with the amperage setting.

Within an electrical circuit there is *resistance*. Resistance is the opposition to

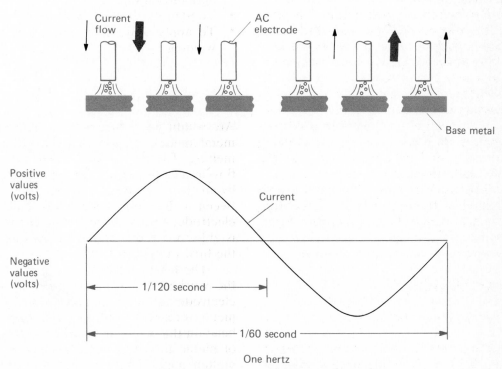

Figure 15-2.   Flow of current in ac welder circuit.

the flow of current in a circuit and causes electrical energy to be transformed into heat. In the arc welder, the distance between the tip of the electrode and the base metal, called the arc length, forms a high resistance to the current flow. This results in the release of a large amount of heat into the welding zone.

Many of the farm-type welders use alternating current (ac). During the welding process, alternating current is constantly reversing *polarity*. The polarity indicates the direction the current is flowing in a circuit. Figure 15-2 illustrates the reversing of the current flow in the arc welder.

Some arc welding processes are done using direct current (dc). The circuit has electrical current flowing in one direction only. When the electrode holder is attached to the negative (−) terminal and the ground cable is attached to the positive (+) terminal, the circuit has straight polarity (DCSP). When the electrode cable is attached to the positive terminal and the ground cable is attached to the negative terminal, the circuit has reversed polarity (DCRP). Figure 15-3 shows how the cables are connected in each case.

Controlling Distortion

In addition to understanding basic electricity, a welder must understand the physical properties of metal. When heated, metal expands in size. As it cools, it contracts, or shrinks to the original size. *Distortion* occurs when the natural expansion or contraction is restricted. Although expansion and contraction cannot be eliminated, they can be controlled to prevent distortion of the welded metal.

Use proper welding methods.   By using proper methods, a welder can counteract the expansion and contraction forces.

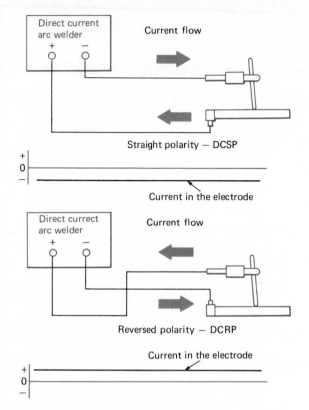

Figure 15-3.   Flow of current in dc welder circuit.

Every joint should be tack welded at each end and at regular intervals along the joint. If the joint is longer than 10 in [25.4 cm], it should be tacked every 10 or 12 in [25.4 or 30.4 cm]. Tack welds should also be placed at points where stresses could change the position of the base metal. As a general rule, the length of the tack weld should be approximately twice the thickness of the base metal. Because of the heat buildup during the welding process, the welder may allow one bead to cool while welding at a different point along the joint.

A back-step method also reduces distortion. Rather than running one long continuous bead, the welder may run a series of short beads (see Figure 15-4).

Other techniques which may aid in reducing distortion include preheating and

BASIC ARC WELDING

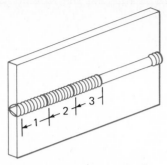

Figure 15-4. The back-step method is sometimes used to reduce metal distortion.

peening. The area around the weld joint may be preheated, causing uniform expansion. As the weld cools, there is also contraction of the area around the joint. Additional shrinkage may be controlled by lightly peening, or striking, the surrounding area with a ball-peen hammer. This technique helps to relieve the internal stress, thus reducing distortion.

Use contraction forces. With some experience, a welder is able to estimate the amount of expansion that will occur during welding and the amount of contraction that will occur as the welded joint cools. By tacking the piece slightly out of position in the opposite direction from the contraction force, the piece will be pulled into alignment as the weld joint cools. Another technique that is used prebends the piece to be welded. As contraction occurs, the piece is unbent into the correct alignment. Also, by welding in the proper sequence, the contraction forces of one weld will be counteracted by the forces of another weld. If possible, weld on opposite sides of the base metal to reduce the distortion of the piece.

Use clamps. In addition to the techniques above, welders often use clamps and jigs to hold the base metal in a fixed position. A jig is a mechanical device used to hold the base metal in a rigid position during welding. Jigs are commonly used in agricultural industry because they allow faster welding and reduce distortion.

Requirements For a Sound Weld

There are many factors affecting weld quality. Most of these fall into four basic groups: (1) using the correct amperage setting, (2) using the correct arc length, (3) using the correct speed of travel, and (4) using the correct electrode position.

Using the correct amperage setting. Distortion can often be traced to the application of too much heat during welding. The amperage (current) required for a particular weld joint depends on the thickness of the base metal and the diameter of the electrode. The thicker the base metal, the more heat is required to maintain a molten puddle necessary for fusion welding. However, when this heat is greater than necessary, excessive expansion occurs causing the weld joint to become distorted. Table 15-1 lists the suggested amperage settings for commonly-used electrodes and metals.

Using the correct arc length. When the arc is started, the distance from the tip of the electrode to the base metal is called the arc length. Like the amperage setting, the arc length depends on the thickness of the base metal and the size of the electrode. As a general rule, use an arc length equal to the diameter of the bare end of the electrode.

The correct arc length not only produces the correct heat and penetration but also produces a gaseous shield around the molten puddle. An arc length that is too short reduces the amount of heat and penetration. Also, there is a greater risk of slag (waste from the flux on the electrode)

TABLE 15-1. Suggested Amperage Settings

METAL THICKNESS (in inches)	ELECTRODE SIZE				
	5/64	3/32	1/8	5/32	3/16
22 gauge	5	20			
1/16 in (1.6 mm)		40			
1/8 in (3.2 mm)		65	75	95	
3/16 in (4.8 mm)		85	105	135	155
1/4 in (6.4 mm)			135	155	175
5/16 in (7.9 mm)			145	175	200
3/8 in (9.5 mm)			165	185	230
1/2 in (12.7 mm)				195	250
Cutting			185	250	

Adapted from Forney Welding Co., Ft. Collins, CO.

being trapped in the molten metal and seriously weakening the weld joint. An arc length that is too long will cause the molten metal to be splashed around the bead. Also, the molten metal may oxidize (change its physical characteristics) because oxygen combines with the liquid metal.

A welder soon recognizes the correct arc length by the sound of the arc. The correct arc length produces a crackling sound similar to frying bacon. An arc which is too long will sputter or snap because of the high resistance to current flow.

Using the correct speed of travel. Because the welder is depositing molten metal into the weld joint, speed is very important. When applying toothpaste to the toothbrush, the speed with which the tube is moved across the bristles affects the width and uniformity of the toothpaste applied. The same principle applies when arc welding. With experience, the welder is able to judge the speed necessary to produce a bead with a width 1½ to 2 times the diameter of the electrode.

Strike the arc and hold the electrode in one spot until a crater twice the diameter of the electrode is formed. Move the electrode forward very slowly while maintaining the correct arc length. If the electrode is moved too fast, the bead will be narrow with skips or breaks. If the electrode is moved too slowly, excessive width and buildup result. This extra reinforcement does not add to the strength of the weld joint. In fact, many times extra buildup results in joint failure.

Using the correct electrode position. The position of the electrode determines the penetration of the molten puddle and controls the slag. For flat position welding, the electrode is tilted 10 to 15° in the direction of travel. This allows the welder to view the weld crater and better judge the width and travel speed. A 75° lead angle allows the arc to dig into the weld joint. At the same time, the slag formed from the flux coating can wash through the molten puddle and form a smooth covering. This slag cover slows the cooling rate and helps to strengthen the weld joint. Figure 15-5 shows both the side and front view of the electrode while running a flat position weld joint.

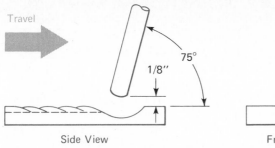

Travel

75°

1/8″

Side View

Lead angle

90°

Front view

Working angle

Figure 15-5. Correct electrode positioning for flat position arc welding.

Evaluating the Weld

The American Welding Society has developed several systems for weld evaluation. These include tests that break, or attempt to break, the weld under pressure, x-ray examinations of the internal characteristics, and measurement of sound waves transmitted through the weld. The most common type of evaluation used in agricultural industry is a visual examination of the weld. This is a good test if the correct factors are examined. After some experience, a welder can visually judge a weld joint and estimate the strength and quality of the weld. A score card is included in the *Agricultural Mechanics Activity Guide and Project Plan Book* to aid in the evaluation.

Width and buildup.   These two factors indicate the strength of the weld. The width of the weld typically should be 1¹/₂ to 2 times the diameter of the electrode used. Seldom should the width exceed ¹/₄ in [6.3 mm] when using a ¹/₈-in electrode. The buildup should produce a smooth convex surface with uniform thickness.

Appearance.   The weld should have a smooth surface with a dense ripple pattern caused by the cooling and hardening of the molten metal.

Face of the bead.   The face, or upper surface, of the bead should be uniformly convex, slightly rounded, with a smooth curved surface. It should have no hills and valleys.

Edge of the bead.   The edge of the bead should appear to be tied into the surface of the base metal without excessive undercutting or overlapping. The edge should be straight, with a bead of uniform width.

Beginning and ending.   This indicates the amount of penetration and the resistance of the weld to tear under stress. The beginning should show full penetration over the tack weld without excessive buildup. The crater at the end of the bead should be filled and smooth.

Surrounding plate area.   The area should be free of weld spatter (small metal droplets). If the wrong arc length was used, the droplets will appear over a wide area. This also may indicate a lack of penetration or that excessive amperage was used.

Slag formation.   The slag should form a smooth coating over the face of the bead. This allows the weld to cool slowly and improves its strength. Some electrodes produce more slag than others and may appear somewhat different. The slag should be easily removed with a chipping hammer and a wire brush. Improper electrode angle may cause the slag to be included in the weld.

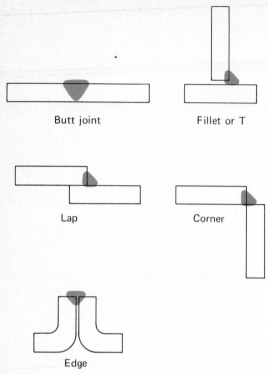

Butt joint

Fillet or T

Lap

Corner

Edge

Figure 15-6.   Five common welding joints.

Types of Weld Joints

There are five basic weld joints used in agricultural industry. The selection of a particular joint depends on the location of the weld, the strength which is required, the requirements of the project, and, to some extent, the time required to prepare the joint.

The five joints commonly used include the butt, fillet or T, lap, corner, and edge. Figure 15-6 illustrates each of the joints. Each joint has certain advantages and limitations. The welder must examine the requirements of the project and select the joint best suited for the job.

Each joint may be welded in one of four positions: flat, horizontal, vertical, and overhead. Because of its most common use, the flat position weld should be mastered first. Figure 15-7 illustrates the four positions.

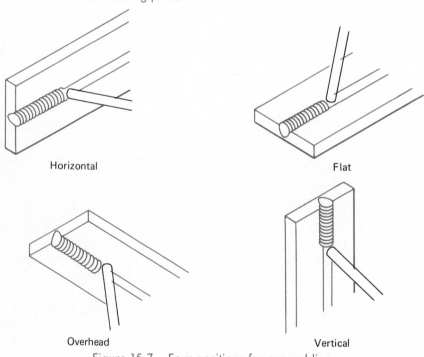

Horizontal

Flat

Overhead

Vertical

Figure 15-7.   Four positions for arc welding.

BASIC ARC WELDING

Basic Safety Practices

Because the worker is using electricity and molten metal, care must be taken to provide a safe work area. Employers are very safety conscious because accidents not only injure skilled workers but cost time and money as well. An agricultural worker must be alert at all times. Agricultural industry is the second most hazardous occupational area in terms of numbers of deaths and cost.

Proper personal protective equipment is a must. A welder must use a welding helmet, leather gloves, and proper clothing. A section in the next chapter discusses these in detail.

Proper ventilation is important because of the amount of smoke and fumes. Several metals such as galvanized steel and chrome produce toxic fumes when heated. Always use a ventilation fan when welding in confined spaces.

Always inspect the equipment. Improperly wired or unprotected welders should not be used. Also, frayed cables or loose clamps should be replaced. Be sure the main disconnect switch is off before making any repairs on the equipment.

THINKING IT THROUGH

1. What is meant by the term *fusion welding*?
2. Match the phrase which best describes each electrical term:

A.	Amperage	1	direction of current
B.	Voltage	2	electrical pressure
C.	Resistance	3	current which reverses direction
D.	Polarity	4	flow of electricity
E.	Alternating current	5	electrode positive
F.	Direct current	6	electrode negative
G.	DCSP	7	opposition to the flow of electricity
H.	DCRP	8	current which flows in one direction

3. List and describe three techniques which reduce or control distortion.
4. Describe the four requirements of a sound weld.
5. What factors should be used in the visual evaluation of a weld?
6. Sketch the five basic weld joints which are used in agricultural industry.
7. Why is safety important when arc welding?

CHAPTER 16

ARC WELDING EQUIPMENT

Before learning how to weld, you need to identify and describe the different types of arc welding equipment. It is also important to be able to operate the equipment safely.

The worker who does not become thoroughly familiar with the arc welding equipment is liable to be injured or injure someone else.

CHAPTER GOALS

In this chapter your goals are:

- To identify the three major types of arc welding equipment.
- To identify the different types of electrodes and determine how they are used.
- To list the rules for safe operation of arc welding equipment.

Types of Arc Welders

There are three basic types of arc welders that the agricultural worker will find common to the agricultural industry. All three are operated by electricity. They are the transformer-type, the engine-driven type, and the rectifier-type arc welders.

The capacity of the three types of arc welders is rated according to the amperage and the duty cycle of the machine. This is based on the percentage of time the arc welder can be operated at maximum amperage during a ten-minute period. A 225-ampere (A) arc welder with a 20 percent duty cycle can be safely operated for two minutes during each ten-minute period at *maximum* amperage. Several manufacturers show the duty cycle at each amperage setting. One machine lists a 100 percent duty cycle at 120 A, a 40 percent duty cycle at 180 A, and a 20 percent duty cycle at 230 A. This means that the machine can be used a full ten minutes at 120 A, four minutes out of ten at 180 A, and two minutes out of each ten-minute period at 230 A. The most common rating method, however, is to list only the maximum amperage setting of the machine.

Alternating-Current (ac) Welders

Most welding equipment used on agricultural jobs uses alternating current (ac). Alternating current means that the current alternates between positive and negative and changes direction 120 times per second. This is also referred to as 60 hertz (Hz). One *hertz* is a unit of frequency equal to one cycle per second.

The typical ac welder used in farm and mechanics' shops may be purchased at a cost of $125 to $400. There are two basic types of ac welders: the transformer and the engine-driven (sometimes called a *generator*).

The Transformer-Type Welder.

The transformer-type arc welder has a transformer built into the welder which increases the amperage (electric current) while decreasing the voltage (electric pressure). These welders are connected to a 240-volt (V) source. The transformer decreases the 240-V source (pressure) to between 15 and 25 V. The arc welder may draw less than 50 amperes (A) from the source. So, the transformer steps up the amperage that is available. The worker sets a constant amperage for a particular job, and the transformer makes the necessary reduction in voltage (see Figure 16-1).

The only adjustment on a transformer-type ac welder is made in the amperage setting. The adjustment may be made with either a crank that raises or lowers an iron core in the coil or a series of tap sockets and plugs. These arc welders are available with maximum amperage ranges of 150–250 A, 250–300 A, and 300–600 A.

The 150–250 A machine is generally considered a limited-input arc welder. This means the machine is not expected to draw over 37.5 A from the 240-V source. This arc welder can be used satisfactorily on rural distribution power lines without causing excessive voltage-drops on the line.

This size arc welder is designed for light to medium welding jobs. It is excellent for farm use when welding is done on a periodic basis.

The 250–300A arc welder is satisfactory for large farm shops and the average needs of agricultural industry. They usually have a 60–80 percent duty cycle and are rugged enough to withstand regular production use. The 300–600A arc welder is designed for heavy-duty welding with large electrodes. They are usually used only in fabricating heavy machine parts, large pipes, and welding thick plates.

The Engine-Driven Welder.

A less common ac welder is engine-driven. It is more portable than other arc welders because it takes a source of power with it. The engine-driven welder is driven with a gasoline or diesel engine and can be mounted on a service truck. As you would expect, the portable arc welder is much more expensive than the conventional welder. A principal advantage of this welder is its ability to produce both ac and dc current.

Direct-Current (dc) Welders

The direct-current welder offers some advantages over the ac welder and is best suited to certain kinds of jobs. The dc welder is similar to the generator on small lawn and garden tractors because it produces di-

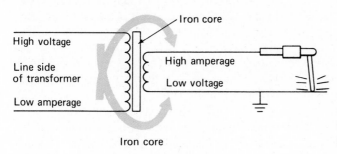

Figure 16-1. A transformer-type arc welder decreases the voltage and increases the amperage.

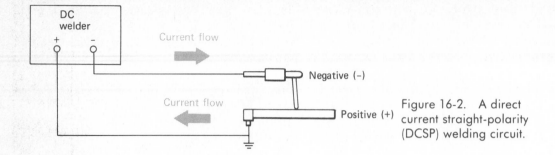

Figure 16-2. A direct current straight-polarity (DCSP) welding circuit.

rect current. The dc welder may produce two types of current: straight polarity and reverse polarity.

Figure 16-2 shows a direct-current welding circuit with straight polarity (DCSP). The electrons, or negatively charged electrical particles, are flowing from the negative terminal—the welding electrode—across the arc and into the base metal and the ground conductor. About two-thirds of the heat produced in this process is released in the base metal while one-third is released in the electrode. This kind of direct current produces a narrow, deep penetrating weld.

The direct-current welding circuit with reverse polarity (DCRP), on the other hand, produces rather shallow weld penetration (see Figure 16-3). In this method, also called *electrode positive*, the current flows from the welding machine through the ground cable and the base metal, across the arc, and into the electrode. From the electrode, the current flows back into the dc welder.

Approximately two-thirds of the heat is given off through the electrode, and one-third is dissipated into the base metal. With the large amount of heat on the electrode, the filler metal and the electrode coating are melted very rapidly resulting in less penetration.

The two most common types of dc welders in agricultural industry are the *engine-driven welder* and the *rectifier-type welder*.

Engine-Driven Welder. The engine-driven welder is portable and is often used in construction work in agricultural industry. The engine drives a large generator that provides the electricity for the welder. The primary drawback to this equipment is that it is more expensive and very noisy.

Rectifier-Type Welder. The rectifier-type arc welder is becoming more common in agricultural shops. The rectifier creates a one-way flow of current using a diode. This principle converts alternating current into direct current. By turning a three-position switch, the rectifier-type arc welder produces both dc with reversed polarity and dc with straight polarity plus ac

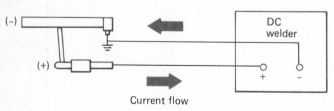

Figure 16-3. A direct current reversed-polarity (DCRP) welding circuit.

BASIC ARC WELDING

current. While the rectifier-type machine is more efficient and much quieter to operate, it lacks the portability of the engine-driven arc welder.

Electrode Identification

The *electrode* used in most welders is a coated steel rod. The rod conducts the electric current that creates the arc and is melted to form the filler metal. The coating is the *flux*. As it burns, it helps clean the base metal, but it also produces an inert gas that protects the molten puddle from contamination by nitrogen and oxygen in the air.

There are a wide variety of electrodes on the market today. The American Welding Society (AWS) has developed a numbering system that aids the worker in selecting the correct electrode for a specific job. Each four-digit number in the classification system is preceded by the letter E, which identifies the electric arc electrode. The first two digits in the number indicate the tensile strength of the weld metal. If the digits are 70, for example, it means the weld metal will have a tensile strength of 70,000 pounds per square inch (psi) [483 MPa]. Most low-carbon or mild steel has a tensile strength of 60,000 to 75,000 psi [414–517.5 MPa]. The third digit identifies the position(s) in which the electrode can be used. The welding positions refer to the plane (flat, vertical, etc.) in which the weld is made. The number 1 indicates that the electrode can be used in all positions. The number 2 indicates the electrode is suitable for flat and horizontal positions only. The number 3 is designed for welds in the flat position only. The fourth digit is a key for several things: the type of current, ac or dc, and whether the electrode may be used as DCSP or DCRP, the type of flux coating, the pene-

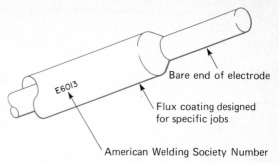

Figure 16-4. Most manufacturers use the AWS number to identify the coated electrode.

tration of the arc, and special applications. See Figure 16-4.

If you selected an electrode numbered E6013, the weld metal has a tensile strength of 60,000 psi [414 MPa], that the electrode can be used in all positions, and that the arc has a stabilizer in the flux coating and produces a medium to shallow penetration.

In addition to the number classification system, there is also an older color coding system. An electrode is identified by color dots located on the coating (see Figure 16-5). Table 16-1 presents the National Electrical Manufacturers Association (NEMA) color markings. Most manufacturers now designate the electrodes using the AWS system.

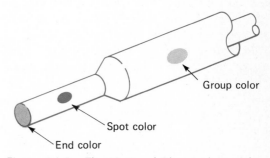

Figure 16-5. The National Electrical Manufacturers Association (NEMA) uses color markings to identify coated electrodes.

TABLE 16-1. NEMA Classification by Color Markings

AWS NO.	GROUP TYPE	GROUP COLOR	END COLOR	SPOT COLOR
E xx11	Mild steel	None	None	Blue
E xx13	Mild steel	None	None	Brown
E xx16	Mild Steel (low hydrogen)	None	None	Orange
E xx18	Low hydrogen (iron powder)	Green	Black	Orange
E 308-16	Stainless steel	Yellow	Yellow	None
ENIFE	Cast iron (ductile)	White	Orange	Brown
ENI	Cast iron	White	Orange	Blue
EST	Cast iron	None	Orange	None
	Hard surface	Red	None	None

Electrodes are sized according to the diameter of the bare end of the rod and the overall length of the rod. The most common sizes used in vocational agriculture are $3/32 \times 12$ in [0.02 × 30.5 cm], $1/8 \times 14$ in [0.03 × 35.6 cm] and $5/32 \times 14$ in [0.04 × 35.6 cm].

Electrode Care

It is important to follow the manufacturer's recommendations regarding the use of electrodes. Each one is designed for a specific job. The electrodes must be stored properly too. If it is too moist where they are kept, the coatings will absorb the moisture. If a damp electrode is used, the heat of the welding operation will turn the water that has been absorbed into steam. The steam, in turn, can cause porosity in the weld metal.

Generally speaking, damp electrodes should be placed in an oven set at 250°F [121°C] for 4 or 5 hours. Many manufacturers have specific times and temperatures for drying their electrodes. To prevent moisture from entering the coating in the first place, electrodes can be stored in the shop in an old refrigerator with a 40-watt (W) bulb. The refrigerator is *not* turned on, but the light is left on as a heat source. Also, lock the door so that small children cannot get into the refrigerator.

The coating of electrodes should not be broken or cracked. Therefore, avoid dropping them or bumping them against sharp edges. When welding, the worker should take only the number of electrodes needed for the job. These electrodes can be kept in a simple, shop-made electrode holder (see Figure 16-6).

MIG Welders

A newer type of welder is the *metal inert gas* (MIG) welder. Sometimes it is referred to by the trade name of the manufacturer such as the micro-wire welder (Hobart), Aircomatic welder (Airco), Sigma welder (Linde), or the Millermatic welder (Miller).

BASIC ARC WELDING

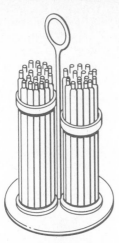

Figure 16-6. A simple electrode holder will prevent damage to the flux coating.

Instead of using conventional electrodes as other arc welders do, the MIG welder uses a small-diameter wire that is fed automatically through a gun to the weld. The most common diameter wire used is 0.035 in [0.89 mm], although other sizes are available. Although most welders use flux-coated electrodes that give off a gaseous substance, the MIG welder uses a gas to shield the weld. The gas used varies with the metal being welded. Carbon dioxide, for example, would be used in welding mild steel, but argon would be used in welding aluminum.

Safety Rules

These safety rules should be followed when you are working with arc welding equipment:

1. Work in a dry location. It is dangerous to use electrical equipment in a wet or damp area.
2. Make sure the equipment is properly grounded to avoid shocks.
3. Gloves should be worn as protection against heat and safety glasses should be worn as protection against intensive light rays.
4. Long-sleeved, fire-resistant clothing should be worn. Cuffless pants and boots also are advised.
5. Pliers or tongs should always be used to handle hot metal.
6. The welding helmet should be worn to protect the head and face from sparks and hot metal chips. The helmet should have a filter plate lens (shade number 10) to protect the eyes against the intense light given off during the welding process.
7. Before you strike the arc to begin welding, call out "cover!" to alert other to shield their eyes or move away.
8. The work area should be well ventilated.
9. Equipment cables should be laid out straight when the welder is being used and hung up properly when the welder is not in use.
10. Know the location of nearby fire extinguishers and how to use them.

Arc Welding Equipment: A Review

The most common arc welder used in production agriculture is the alternating-current welder. Usually, the welder has an amperage range of 225 to 250 A. There are transformer and engine-driven types of ac welder. The dc welder is used more by agribusiness because it can weld more kinds of metal. The rectifier type of dc welder is most common in agribusiness, usually with the dual-purpose (ac/dc) welder. The engine-driven dc welder is preferred for portable use.

Electrodes are used in arc welding, and there are many varieties. The most common electrodes are the 60 and 70 series, which have a tensile strength of 60,000 and 70,000 psi [414–483 MPa].

The metal inert gas (MIG) welder is often used where a great deal of welding is done. The procedure is discussed in more advanced agricultural mechanics courses.

In using any welding equipment, it is vital that the operator be properly protected from both heat and intense light. The worker also must be thoroughly familiar with the equipment and what it can do.

THINKING IT THROUGH

1. What are the two basic types of alternating-current (ac) welders? What are the advantages of each?
2. What are the advantages of a direct-current (dc) welder?
3. What is meant by *polarity*? What are the two polarity settings on a dc welder?
4. How do you adjust the "heat" of the welding process?
5. What are the tensile strength and the possible positions for the following electrodes?
   (*a*) E6011
   (*b*) E7024
   (*c*) E6013
6. How should electrodes be stored?
7. Explain how the MIG welder differs from the conventional electrode welder?
8. What kinds of personal protection equipment should be worn in arc welding?

CHAPTER 17

FLAT-POSITION ARC WELDING

During the arc welding process, metal is superheated by the electric current passing through the weld joint and a small puddle of liquid metal is produced. The puddle of liquid metal has an internal temperature of approximately 6000°F [3316°C]. As the puddle is moved along by the operator, it forms what is called a *bead*.

Since the metal is in a liquid state, it may drip out of the weld joint if it is not controlled properly. A simple way to control the puddle is to weld in the flat position. The piece of metal being welded is positioned so that the weld is made from the top. This is the best position to use while you are learning to weld since it is not necessary to concentrate on controlling the effects of gravity.

Most of the weld joints are made in the flat position. Even an expert will weld in the flat position when possible because of the improved quality of the weld joint.

CHAPTER GOALS

In this chapter your goals are:

- To demonstrate how to correctly start and stop a bead
- To run a bead and build a welding pad
- To be able to weld a butt, fillet, and lap joint in the flat position

Starting and Stopping the Bead

As one learns how to use the electrode, it helps to concentrate on starting and stopping the bead. Then it is easier to control the molten puddle and run the bead correctly.

A $1/8$-in [3.2 mm] E6011 or E6013 electrode should be selected. These electrodes are commonly used in school and agricultural shops. The ground clamp is attached to the welding table or to the metal to be welded. If the metal is not properly grounded, poor welding will result.

For practice, select a piece of scrap steel plate. Mild-steel plates $1/8$ in [3.2 mm] thick and about 3×8 in [7.6 × 20.3 cm] are the easiest to weld.

The amperage control on the welder is adjusted to the 100A setting. This setting will provide a basis to experiment with higher and lower settings and observe the effects. The welding cables should be laid out straight, and the arc welder should be turned on. Place the bare end of the electrode in the holder at a 90° angle to the holder (straight up and down). Suitable protective clothing, including gloves and a helmet with a number 10 filter plate lens, must be worn. Position the electrode near the base metal and call out "cover!" This alerts other people nearby that the arc is about to be struck.

The right-handed person should hold the electrode holder in the right hand. Support the right hand with the left hand. The right arm should be positioned so that the elbow is tucked in close to the body near the hip. Practice positioning the arm until it is comfortable.

To strike the arc, position the electrode so that it is at a 75° angle to the base metal (see Figure 17-1) and toward the direction in which the bead will be moved. Start (strike) the arc by either scratching the electrode (similar to striking a kitchen match) or tapping it lightly on the base metal.

Keep the arc length at about $1/16$ to $1/8$ in [1.6 to 3.2 mm]. Decrease the arc length slightly as the base metal begins to melt. Following a straight line (which can be drawn with soapstone), move the electrode forward slowly. Run a short bead of approximately 1 in [25.4 mm]. This will help you get the feel of the electrode and develop control.

To stop the bead, make a circular motion with the electrode and then quickly move the electrode away from the base metal. This breaks the arc between the base metal and the electrode without excessive spatter and fills the crater of the bead.

After the initial practice at running a bead, stop to review the technique.

Evaluating the Work

When the arc was struck, a flow of electric current between the electrode and the base metal was started. As the electricity passed across the gap—the arc length—a large amount of heat was given off. This heat (up to 12000°F, or 6650°C) affected both the end of the electrode and the base metal. The result was a molten puddle which formed the weld joint.

The high temperature of the electrode tip and the base metal formed a crater of liquid metal (see Figure 17-2). The tip of the electrode should have been kept about $1/16$ to $1/8$ in [1.6 to 3.2 mm] from the crater and the electrode moved in a straight line to form a bead $1/4$ to $3/8$ in [6.4 to 9.5 mm] wide.

After a short bead was run, the weld should have been allowed to cool to a black color. During the welding process, a coat-

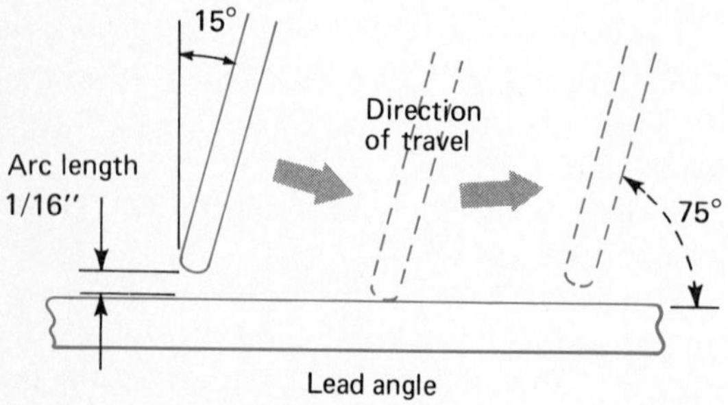

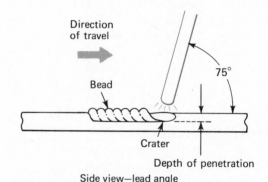

Figure 17-1.   Strike the arc with a scratching motion.

Figure 17-2.   Striking the arc and running the bead.

BASIC ARC WELDING

ing formed on the weld. This is called *slag*. When the weld cools, the slag becomes hard. To remove the slag, first brush the toe or weld with a wire brush. Then, using a chipping hammer, tap the slag lightly near the edge of the bead to loosen the slag. Be sure to wear protective glasses during the brushing and the chipping. The slag must be removed if a second weld is to be applied over the first bead. Otherwise, the slag will be remelted and mixed into the next weld. This will cause a weak weld.

Inspect the short beads using the evaluation factors in Chapter 15. When the starting and stopping of a bead has been mastered, it is time to try running a full bead.

Running a Bead

The *flat bead* is the basic weld. There are four essential requirements to running a correct bead: selection of the correct amperage setting, maintaining the correct arc length, maintaining the right speed, and using the correct electrode angle. One has to be able to do all these things correctly before a good flat weld can be made.

In preparation for running a bead, select a mild-steel plate and draw straight lines on it with a soapstone. The soapstone does not burn off as fast as chalk. Using the 1/8-in [3.2 mm] E6011 or E6013 electrode, place it correctly in the holder (straight up and down). The right-handed person should grasp the holder in the right hand and support that hand with the left hand. Call "cover!" and lower the helmet with a nod of the head. Use a scratching motion to start the arc.

Maintain a 1/16- to 1/8-in [1.6- to 3.2-mm] arc length holding the electrode in one place to build the bead until it is about 3/8 in [9.5 mm] wide and approximately 1/8 in [3.2 mm] high.

1/16" forward per motion

Figure 17-3. Weaving motions used for multipass welds.

Now, move the tip of the electrode forward slowly while maintaining the correct electrode length. Watch the action of the molten puddle very carefully. By watching the puddle, one can better judge the travel speed necessary to produce a uniform bead 2–2½ times the diameter of the electrode (see Figure 17-3).

To evaluate the bead, see whether there is a uniform width and thickness to the bead, whether it has smooth and uniform ripples, whether it is free of dips and high spots, and whether the bead has fused to the base metal without overlapping or undercutting. Also, the crater should be filled. Examine the area for spatter, which indicates that the arc was too long. The slag formation should be uniform and easily removed.

Building a Pad

Sometimes one can preserve the life of parts and equipment by building up worn spots in the metal rather than buying new parts or equipment. This buildup, or padding, can save time and money.

To build a *pad*, first clean the base metal using a wire brush or a grinder. Then lay straight beads side by side, overlapping the edges by about $^1/_{16}$ in [1.6 mm]. Be sure one bead has cooled and all slag and flux have been removed before applying another bead. Otherwise, you risk having a porous weld because slag becomes trapped in the weld.

Applying Second and Third Layers

When the first layer of beads has cooled and has been properly cleaned, it is time to lay the second layer of beads. These beads are laid at right angles to those in the first layer. Again, make sure each bead cools before overlapping another one. The metal being welded is overheated if it turns a dull-red color. Let it cool until it can be touched for 10 or 15 seconds. If the metal is overheated, the structural quality of the metal will change resulting in a brittle weld.

A third layer can be applied over the second layer in the same manner. Remember, the third layer will be at a right angle to the second layer, and it will run in the same direction as the first layer.

To review some important points, remember to overlap the edges of the beads about $^1/_{16}$ in [1.6 mm] and remove all slag and flux from one bead before overlapping another. Brush the surface with a wire brush until it is cool. Keep the bead straight and at a uniform height. Stop frequently to allow the metal and the pad to cool. Each layer is always at a right angle to the layer below. One can apply as many layers as necessary to achieve the desired thickness of the pad.

Flat Butt Weld

One of the most common welds is the flat butt weld. It is used frequently to repair equipment and to construct machinery. Two pieces of metal in the same plane are brought together so that one piece aligns perfectly with the other piece. A butt weld may be placed on only one side of the base metal or the joint may be welded from both sides.

Butt-Welding Sheet Metal. As always, metal preparation is important. Sheet metal with a thickness of less than $^1/_8$ in [3.2 mm] must have all the paint, rust, and foreign material removed with a wire brush. When fitting the two pieces of metal together, allow a gap, or root opening, equal to at least one-half the thickness of the base metal. Tack the base metal at each end and every 8 to 10 in [20.3 to 25.4 cm] along the joint (see Figure 17-4). This will help keep the metal from distorting from the heat.

An E6013 electrode is excellent for butt-welding sheet metal. It does not burn through the sheet metal as easily as some others. Consult Table 16-1 when selecting the electrode for butt welding.

Butt-Welding Metal Plate. In using the butt weld to repair metal that is thicker than $^1/_4$ in [6.4 mm], a backup strip behind the butt weld may be used. A *backup strip* prevents the molten metal from burning through to the worktable and also serves as a reinforcement to the joint. If the metal backup strip is to be a part of the welded

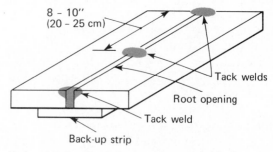

Figure 17-4. The joint must be tacked before the weld is made.

BASIC ARC WELDING

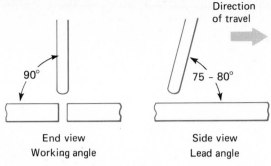

Figure 17-5. Positioning the electrode for the flat butt joint.

joint, use a strip of mild steel. If the backup strip is used strictly to prevent burn-through, select a dissimilar metal such as a strip of copper or aluminum.

Procedures for Flat Butt Welding.   If you are right-handed, the weld joint should be positioned at a right angle to the body. Now, with the electrode holder in your right hand, tack-weld the right end of the joint. The tack should be between $^3/_8$ and $^1/_2$ in [9.5 and 12.7 mm] long.

Begin the butt weld on the left end of the base metal. Hold the electrode straight up and down and then tip it 10° to 15° in the direction of the travel as the arc is struck (see Figure 17-5).

Fuse the bead to the original tack weld, and move the bead slowly to the right. Carefully watch the puddle and apply filler metal at the edge of the crater. The molten metal should form a dense ripple pattern across the weld reinforcement. The bead should be approximately $^3/_8$ in [9.5 mm] wide with a straight edge.

In welding sheet metal it is possible to melt through the base metal. This usually happens if one runs the bead too slowly or sets the amperage too high. If the base metal is melted, stop the weld and allow the base metal to cool. Then, reduce the amperage and restart the arc. Weld a circu-

lar bead around the inside of the hole. Break the arc again to allow that bead to cool. Brush and chip the slag. Restart the arc again, and apply another bead around the inside of the hole. Continue this procedure until the hole is filled. The repaired area will never be as strong as the rest of the joint because some slag is trapped in the hole. After the hole is filled, set the correct amperage and continue the butt weld.

To review the important points, select the correct size and classification of electrode. Be sure to tack the base metal properly, leaving a root opening equal to the diameter of the electrode. The base metal is tacked along the joint, and a backup strip may be used. Increase the speed of travel slightly toward the end, since the metal will be preheated ahead of the bead.

Multipass Butt Welds.   When butt-welding on metal that is thicker than $^1/_4$ in [6.4 mm], bevel the edges before starting the weld. This ensures full penetration of the joint.

When the base metal is thicker than $^1/_8$ in [3.2 mm], the joint may require a multiple-pass weld (see Figure 17-6). The first bead is called a *root pass*, and it is applied to the beveled joint just as if one were welding sheet metal. The root pass should fill the root opening and fully penetrate the joint. After the first bead has been made, allow it to cool. Brushing with a wire brush will speed up the cooling process and help to remove the slag. Chip the slag and inspect the bead. Look at the back of the

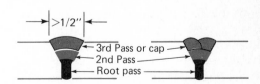

Figure 17-6. Multiple-pass butt weld.

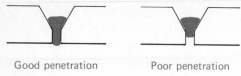

Good penetration   Poor penetration

Figure 17-7.   Penetration of the root pass.

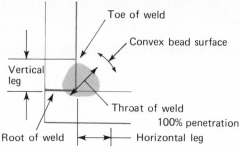

Figure 17-8.   Parts of the fillet weld.

joint to be sure that the first bead has fully penetrated (see Figure 17-7).

Now, deposit the second bead, called the *filler pass*, directly on top of the first one. A slightly higher amperage setting may be needed for the second bead. Make sure, however, that the second bead joins both edges of the base metal. Use a crescent weave (see Figure 17-3) to tie the joint together.

After completing the filler pass, brush the bead to cool it and remove the slag. Rebrush and inspect the bead. The bead should be smooth with uniform ripples and free from slag pockets.

The top bead, often called the *cover pass* or *cap weld*, can now be deposited. This final bead should join the base metal with a smooth, rounded reinforcement. The cap weld should not be more than $1/2$ in [12.7 mm] wide. If it is absolutely necessary to make the cap weld wider, use two beads.

In making the final evaluation of the butt weld, check the width and appearance of the weld. It should be smooth, with uniform ripples. The weld should be slightly rounded and not extend more than $1/16$ in [1.6 mm] above the base metal. The beads should fuse well with the base metal with no overlapping or undercutting on the margin. The crater should be well filled at both the beginning and the end of the joint. Finally, the base metal should be free of spattered metal, and the slag should be completely removed. A good bead does not require grinding unless it is necessary for clearance of another part.

Fillet Weld

A *fillet* (''fill-it'') *weld* joins two pieces of metal at a right angle to each other (see Figure 17-8). Many principles of butt welding can be applied.

The fillet weld not only fills the joint but also extends up the vertical piece of metal and across the horizontal piece. These extensions of the weld are called the *legs*. The leg generally should be as long as the base metal is thick. If the two pieces of the base metal are of different thicknesses, the leg should be equivalent to the thickness of each piece.

When setting up the joint, allow a root opening of about $1/8$ in [3.2 mm]. This is done by tacking the pieces along the joint as with the butt weld.

Running a Multipass Fillet.   When you make the root pass for the fillet weld, tip the electrode approximately 15° in the direction of travel and at a 45° angle to the horizontal piece of metal, bisecting the joint. Begin the bead by fusing to the tack weld. As the bead is run, maintain an arc length of $1/8$ in [3.2 mm] without any weaving motion. Be careful that the vertical leg of the bead does not undercut or overlap.

If the base metals are of different thicknesses, concentrate most of the heat of the filler pass on the thicker piece.

BASIC ARC WELDING

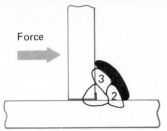

Excess reinforcement pries the root of
the joint up - - causing weld failure

Figure 17-9.   Fillet weld with extensions.

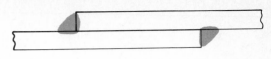

Lap weld

Figure 17-10.   Lap joints are commonly used to
repair agricultural machinery.

Watch the crater carefully to make sure the molten metal is flowing into the joint.

If the cap weld is too slow, the bead may pile up. This will act as a wedge if the force on the welded piece is applied against the root of the bead (see Figure 17-9). If the travel speed of the bead is run too fast, the bead may be too narrow and have a groove down the center.

After the fillet weld is completed, brush it to cool the joint and help relieve the stress. Inspect the joint to be sure that the weld is slightly convex and free of dips and high spots. There should be no slag on the weld, and both legs should be about the same width. The edges of the bead should fuse well with the base metal, and the beginning and end of the joint should be well filled.

Check the back of the weld joint to see if there is full penetration. A section of the fillet weld should be cut with a saw so that the quality of the welded joint can be examined.

Lap Weld

A lap weld (see Figure 17-10) is applied in much the same way as the fillet weld. The lap weld is used when added strength is required and the two pieces of metal can be laid over each other. A farm equipment mechanic or a farmer may use a lap weld to join the heavy frames on a livestock trailer.

Usually, both sides of the lap are welded to prevent bending. To begin, overlap the two pieces by at least 2 in [5.1 cm], and tack-weld on both weld joints. Hold the electrode at a 60° angle in the direction of travel and at a 45° angle to the horizontal leg of the weld joint.

As with the other welds, be sure the bead is fused with the tack weld on the end before you move forward. Move forward at a uniform speed, and tie the edges of the weld to the top edge of the overlapping piece. Make sure the filler metal fuses properly with both pieces of the base metal.

If the two pieces just joined are more than $3/8$ in [9.5 mm], a multiple-pass lap weld should be made. Use the same technique as with the multiple pass on the butt or fillet weld. But make sure each bead is free from slag before applying the next one (see Figure 17-11).

Safety Rules

These safety rules should be followed when you are welding in the flat position:

1.   Wear protective clothing and the welder's helmet with a filter plate lens.
2.   Call out "cover!" when striking the arc.
3.   Remember to use tongs when handling hot metal.
4.   Don't let sparks or melted metal get near flammable materials.

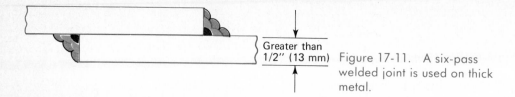

Greater than
1/2" (13 mm)

Figure 17-11. A six-pass welded joint is used on thick metal.

Flat-Position Arc Welding: A Review

When given a choice, most welders prefer to use the flat position for welding. The bead is easier to run and is usually of better quality.

In learning how to weld, it is a good idea to begin by practicing starting and stopping the bead. The electrode should be held at a 90° angle to the surface and tipped approximately 15° in the direction of travel. An arc length of $1/16$ to $1/8$ in [1.6 to 3.2 mm] should be maintained. Stop during the process to brush the bead, thereby re-moving the heat stress and removing slag.

The three most common joints used in agricultural industry are: the butt weld, fillet weld, and lap weld. Each joint has a particular application and advantage. The metal is joined by using a single-pass bead or a multi-pass series of beads. The first bead, called the root pass, completely penetrates the joint and serves as the base for the other beads. The second pass, called the filler pass, ties the two pieces of metal together. The cap weld is the last bead. It ties the edges of the base metal into a single piece.

THINKING IT THROUGH

1. Make two sketches of the flat butt weld, showing the proper position of the electrode in relation to the base metal. Show the side and front views.
2. Describe why it is important to remove the slag before adding another bead to the welded joint.
3. What preparatory steps should be followed for a butt weld on materials that are $1/4$ in [6.4 mm] thick?
4. What determines the leg lengths of a fillet weld?
5. What is the primary advantage of a lap weld compared to a butt weld?

CHAPTER 18

OUT-OF-POSITION WELDING AND SPECIAL USES

Although the welder always prefers to weld in the flat position, sometimes it is impossible to weld in the flat position.

For example, a piece on a heavy disk may break while plowing a field. Not only does the repair on the equipment have to be made in the field, but it probably will have to be made out of position.

The out-of-position weld includes any weld that is made in a position other than the flat position. Out-of-position welds include the horizontal, vertical, and overhead welds. Although horizontal-, vertical-, and overhead-position welding are more difficult than flat-position welding, many of the same basic welding techniques still apply.

Occasionally, an agricultural worker must make repairs without the benefit of a large, well-equipped shop. A logger or land management worker, for example, may have to make repairs to equipment at the site. In such cases, the workers must rely on portable equipment. And this equipment may have to be used to cut a replacement part or to pierce new holes.

Portable arc welding equipment can be used both to cut metal and to pierce holes.

A farm or ranch worker may be disassembling a rusty machine when a stud bolt breaks off flush with the threads. If the bolt cannot be removed with a screw extractor or by other means described in Chapter 12, it may be necessary to weld a nut to the top of the bolt.

Another special arc welding job that may be performed during the off season is to hard-face plowshares, chisel sweeps, and other tillage tools. Parts with hard-facing have a much greater life span than those without. Cutting edges on tillage tools stay sharper longer, reducing power requirements and increasing speed.

While many workers will not have to do out-of-position welding or other special jobs, the fact that they know how to do this kind of work makes them that much more valuable to employers in agricultural industry. For example, if the employer can occasionally call on an employee who possesses out-of-position welding skills, it will save the employer the time and cost of getting a specialist to do the job on site or in the mechanic's shop.

CHAPTER GOALS

In this chapter your goals are:

- To complete a horizontal bead and the butt, lap and fillet weld joints
- To complete a vertical-up bead and the butt, lap and fillet weld joint
- To complete an overhead bead and a butt-weld joint
- To weld a pipe using a butt-weld joint
- To select electrodes that are used to cut mild steel with the arc welder
- To demonstrate how to pierce a hole using the arc welder
- To discuss and demonstrate the technique of removing a broken stud bolt
- To select the proper electrodes and demonstrate the procedures used to hard-face parts with an arc welder
- To demonstrate the safety rules that protect both you and others from injury

Welding in Different Positions

The flat position was discussed in the previous chapter. The three other positions are *horizontal*, *vertical*, and *overhead*. While in these positions, the worker can weld the butt, fillet, and lap joints.

Horizontal Welding

A horizontal weld differs from a flat-position weld because it is made in the horizontal position on a vertical surface (see Figure 18-1). When applying this weld, the welder must be able to control the molten puddle and prevent it from flowing naturally down and away from the joint. To prevent this, one uses an arc length that is shorter and an amperage that is slightly lower than a similar weld in a flat position.

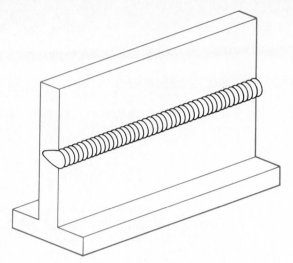

Figure 18-1.   A horizontal butt weld.

Also, an electrode should be selected that has *fast-freeze characteristics*. This simply means that the filler metal from the electrode solidifies faster than the filler metal from many other electrodes.

A common problem with horizontal welds is an overlapping of the filler metal on the lower piece and undercutting on the upper piece. This can be corrected by using the proper electrodes and techniques (see Figure 18-2).

Procedures For Horizontal Welding.   Begin by holding the electrode at an angle of about 75° or 80° to the lower plate (see Figure 18-3). This angle will help

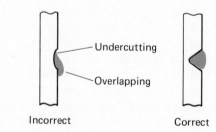

Figure 18-2.   A common horizontal welding problem.

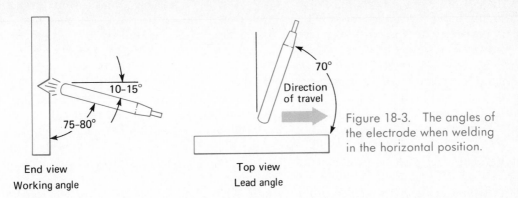

Figure 18-3. The angles of the electrode when welding in the horizontal position.

End view
Working angle

Top view
Lead angle

direct the filler metal up slightly and fill the top edge of the bead. Also, tip the electrode about 20° in the direction of travel.

Use a narrow crescent motion to reduce the dripping of the molten metal. Keeping the arc length at approximately $\frac{1}{16}$ in [1.6 mm], move the tip of the electrode across the crater. Pause slightly at the end of each motion to allow the molten puddle to solidify. Move forward approximately $\frac{1}{16}$ in with each crescent motion.

When applying a multi-pass horizontal butt weld, remove all slag from the root (first) pass before applying the second bead. Also, only the top piece of the base metal should be beveled. The edge of the lower piece, then, becomes a good shelf for the bead (see Figure 18-4).

Running the Horizontal Lap Joint. A horizontal lap is made much like the fillet

weld because what is essentially a fillet bead is applied to the upper joint (see Figure 18-5). Hold the electrode at about a 45° angle to the top piece of metal, and angle the electrode 15° in the direction of travel. The profile of the bead should be like other beads—slightly convex with the vertical leg of the weld equal to the thickness of metal.

The lower horizontal lap weld requires a slightly different technique. Because of its position, this weld is similar to the overhead position. The electrode should form a 30° angle with the surface of the lower piece and be tipped 10 to 15° in the direction of travel (see Figure 18-5).

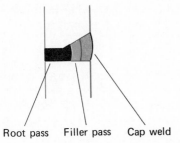

Root pass   Filler pass   Cap weld

Figure 18-4. A multi-pass horizontal butt weld.

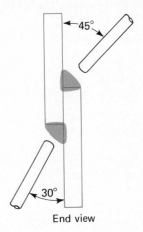

End view

Figure 18-5. Welding a horizontal lap joint.

The arc length should be shorter than it was when the upper lap weld was done.

In evaluating the horizontal weld, look for the same factors as on other welds: width and buildup, appearance, full penetration, slag, excessive spatter, and beginning and ending of the bead.

Vertical-Down Welds

A *vertical-down weld*, sometimes referred to as a *downhill weld*, is done on a vertical joint on a vertical surface. The weld begins at the top of the joint and is welded toward the bottom. This weld is excellent to join light-gauge metal because of the shallow penetration. It is important to know that the vertical-down weld does not have the penetration of many other welds. Also, the weld must be applied using a faster travel speed.

Procedures for Welding Vertical-Down. To begin a weld on ¹/₈-in [3.2 mm] metal or metal that is thinner, tack both ends of the joint and at every 6 to 8 in [15.2 to 20.3 cm]. Starting at the top of the joint, point the electrode up at an angle of about 60° to the surface. Use a short arc and keep the crater size minimal.

Travel should be just fast enough to keep the molten puddle behind the arc. Use a very slight weaving motion to give the bead the correct width (see Figure 18-6).

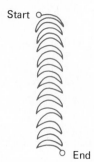

Start

End

Figure 18-6.   Running a vertical-down bead.

The *vertical-down butt weld* is formed much like the bead. Tack the pieces and use a slight weaving motion to produce the desired width of the bead. Hold the electrode at a right angle to the vertical surface. As the work proceeds, one will see that the penetration is shallow, probably less than one-half the thickness of the metal.

The *vertical-down lap weld* is a modified fillet. The electrode should be held at a 45° angle to the surface of the piece, dividing (bisecting) the angle of the joint, and with the electrode tipped slightly downhill (5° to 10°). Use the crescent motion to tie the edges of the pieces and produce the correct leg length. The face of the weld should not show signs of dripping— what some welders call *icicles*.

The *vertical-down fillet* is a fairly easy weld to control. The electrode is held at a 45° angle to the surface of the top piece, bisecting the angle of the joint, and is tipped 5° to 10° in the direction of travel. Do not use a weaving motion. Instead, pull the electrode straight down the joint, traveling just fast enough to stay below the molten puddle. Many welders prefer the E6010 or E6013 electrodes because of their fast-freeze characteristics. Check the technical manuals before selecting the electrode.

Vertical-Up Welds

A *vertical-up*, or *uphill weld* is done on a vertical joint on a vertical surface. This weld has deep penetration and should be used on metal that is more than ¹/₈ in [3.2 mm] thick and where maximum strength is desired.

Procedures for Vertical-Up Welds.   As in other welding operations, prepare the metal first. Grind a 30° bevel along the edges of the metal. Tack the joint as on other welds.

Master the vertical-up bead before at-

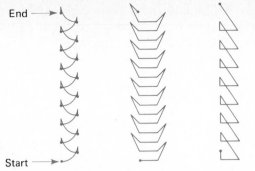

Figure 18-7. Three whipping motions used when running a vertical-up weld.

tempting a complete weld. Beginning at the bottom of the joint, create a small shelf with the initial bead. Move the tip of the electrode up in a short whiplike motion. That is, after having deposited molten metal, move the electrode quickly forward about a 1/2 in [12.7 mm] without breaking the arc. This allows the deposit to solidify before another deposit is made on top. Use this whipping motion all the way to the top of the bead (see Figure 18-7).

A *vertical-up butt weld* should have a root opening of approximately 1/8 in [3.2 mm] or half the thickness of the base metal, whichever is less. On multiple-pass beads, the root pass should be applied without the whipping motion. This pass should penetrate to form a small bead on the back side of the joint. It may be necessary to use a backup strip to prevent burn-through.

As in other welds, be sure to clean away slag from one bead before applying another. The filler pass should use one of the weaving motions and produce a full-width bead (see Figure 18-8). The cap should be applied as the last pass. It will require concentration and practice before one is able to produce a quality vertical-up weld.

The *vertical-up lap weld* is made by starting at the bottom of the joint. The electrode should be held at a right angle to the metal, bisecting the angle of the joint (see Figure 18-9). Use the same whipping motion as for other vertical welds, but be careful not to increase the arc length and the arc.

The *vertical-up fillet* is accomplished in much the same way as the lap. Hold the electrode at almost a right angle to the metal, again bisecting the angle of the joint. The tip of the electrode should be pointed slightly up. Use a slight whipping motion for best results.

Overhead Welds

Welding in the overhead position requires some special personal protection. Cuffless pants and boots are always recommended. The worker also will need a welder's cap to keep sparks off the top of the head and special leather sleeves that prevent molten

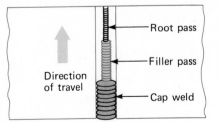

Figure 18-8. A three-pass vertical-up butt weld.

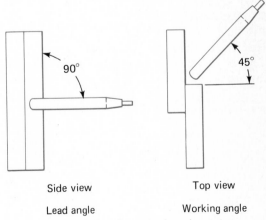

Figure 18-9. A vertical-up lap joint.

metal from falling down on the arm. In addition to these precautions, take another look at the location of fire extinguishers—just to be safe.

In welding in a standing position, it may be more comfortable to drape the welding cable over the shoulder. This will eliminate holding the full weight of the cable during welding.

Depending on the position in relation to the joint being welded, the electrode may be held straight out from the holder rather than at a right angle to the holder. As always, prepare the metal correctly, including tack welding. Select an electrode with fast-freeze characteristics. An E6011 electrode usually works well. Reduce the amperage slightly. To begin the weld, hold the electrode at a 90° angle to the metal, tipping it slightly toward the direction of travel (see Figure 18-10). This is one of the more difficult welding joints, and some time will be spent mastering the techniques in advanced classes.

Pipe Welding

As referred to here, *pipe* is used in construction as a support member rather than to transport water or other materials.

Regardless of how the weld is performed, some of the work will be done out of position. Special care must be taken if the pipe has a galvanized coating because the zinc fumes given off are toxic. Therefore, make sure there is proper ventilation.

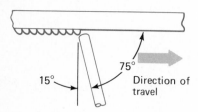

Figure 18-10. Welding in the overhead position.

If the ends of the two pipes are to be joined, the weld is similar to a butt weld. Bevel the ends of the pipe at a 30° angle, leaving a $1/16$-in [1.6-mm] shoulder. Lay the pipe sections in a 2×2 in [5.1 × 5.1 cm] angle iron, and tack-weld at four points. Leave a $3/32$-in [2.4-mm] root opening to ensure full penetration. Chip and brush the slag from all tack welds.

If the pipe sections can be rotated, hold the electrode at a right angle to the pipe joint and weld one-fourth of the joint. Turn the pipe over and weld the opposite quarter. Brush and chip the slag from both beads. Complete the weld between the two quarters. Brush and remove slag from all welds and inspect for penetration.

Safety Rules

These safety rules should be followed when you are welding out of position:

1. In addition to wearing the standard protective clothing, the top of the head and the arms need special protection in overhead welding.
2. If the pipe has a galvanized coating, guard against inhaling fumes given off by the burning zinc in the coating. The area must be well ventilated.
3. All other safety rules described in previous chapters apply.

Cutting with the Arc Welder

By using an electrode at a high amperage setting, it is possible to use the arc welder to cut mild steel and cast iron. Unlike using the oxy-fuel cutting torch (described in Unit VI), the arc welder does not burn through the metal. Rather, the metal melts, and the molten metal runs out of the cut. Consequently, the agricultural worker needs to take special precautions to guard against fire that may result from the droplets of molten metal. Always cut over a suitable container (one that cannot burn). Even a concrete floor will be damaged by molten metal.

The cut made by the arc welder is not as smooth as that made by the oxy-fuel torch, but the cut may be ground smooth if necessary.

A large amount of smoke is given off in the cutting process. Make sure there is proper ventilation in the work area. Galvanized metal produces zinc fumes that are particularly toxic and can cause severe headaches.

Before cutting up a barrel or metal cylinder, find out what was in it. Do not cut into any container that has been used for flammable materials. People have been killed while cutting barrels that were used for gasoline or other combustible materials. As a precaution, cut at least half of the top away using a cold chisel or a saber saw. Also, the barrel must be filled with water to help prevent an explosion.

Select an electrode with deep penetration characteristics. Most operators prefer an E6011 electrode with a diameter of $1/8$ in [3.2 mm] and use an amperage of 180 A to 225 A. Do not set the amperage so high that the electrode becomes red-hot. Some operators dampen the electrode with water before cutting. This causes the coating (flux) to burn more slowly, giving more direction to the filler metal and producing a faster cut.

To cut mild steel, begin by marking the location of the cut with soapstone. To assist in keeping the cut straight, clamp an angle iron along the soapstone line. Always have the base metal in a flat position so the molten metal can fall away from the cut. Start the cut by holding the electrode at a 60° angle to the line made by the soapstone, and strike the arc. Hold the electrode in this position until a deep crater is obtained. Maintaining a very short arc,

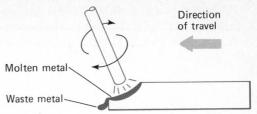

Figure 18-11. Cutting can be done with an arc welder.

move the electrode forward and backward over the cut area, as if wiping the cut. As this "wiping" takes place with the arc, drag the molten metal out of the cut and allow it to fall in the container under the work area (see Figure 18-11).

Keep the area of the cut between $1/4$ and $3/8$ in [6.4 to 9.5 mm] wide. This allows the molten metal to flow out freely. By using a rapid up-and-down motion (wiping), the cut will be smoother and free of excess slag.

Piercing Metal with the Arc Welder

To pierce a hole in metal with the arc welder, hold the electrode directly over the metal and strike the arc. Use a small circular motion (see Figure 18-12) to create a molten crater. Maintain a short arc, keeping the tip of the electrode in the crater until the hole has cut through. To enlarge the hole, use a wider circular motion to dig out the molten metal.

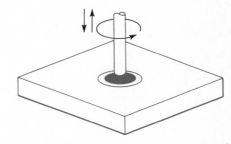

Figure 18-12. Piercing a hole with an arc welder.

When piercing metal thicker than ¼ in [6.4 mm], start the hole from the bottom of the metal. Use extreme caution to prevent being burned by the molten metal.

Removing Broken Stud Bolts

If a stud bolt has broken off and there is no other way to get it out, the arc welder can be used. Select an electrode with a ¹/₈-in [3.2-mm] diameter, preferably an E6011. Set the amperage control at 115 A.

Select a square nut one size larger than the stud bolt. A ³/₈-in [9.5 mm] stud, for example, would use a ⁷/₁₆-in [11.1-mm] square nut. Position the nut directly over the broken stud. Holding the electrode directly over the stud, lower it through the nut and strike the arc on top of the broken stud.

Do not use a circular motion. Fill the cavity left by the broken stud with molten metal (see Figure 18-13). Then use a small circular motion to weld into the threads of the square nut. Completely fill the nut opening with the filler metal.

Allow the stud and nut to cool slightly. Usually, the heat from the weld will expand the stud, and the cooling will contract it. It should now be possible to remove the stud with a wrench. *It is advisable to practice on scrap metal before trying to remove a stud from an expensive piece of equipment.*

Hard-Facing

When metal parts and tools are subjected to abrasion, it is often desirable to put on a thin surface of a metal alloy to protect the base metal and give it a longer life.

This process is called *hard-facing* or *hard-surfacing*. It is usually done to cutting parts such as plowshares, chisel teeth, disks, and knives.

Many different electrodes are used for hard-facing. Each has special characteristics designed for specific jobs. The electrodes are divided into three basic groups: (1) severe abrasion-resistant, (2) moderate abrasion- and impact-resistant, and (3) severe impact- and moderate abrasion-resistant. It is best to know how the part will be used before selecting an electrode. A severe abrasion electrode, for example, is generally good for such things as plowshares, cultivator sweeps, and similar implements that are subject to much wear but little impact. However, if the plowshare is going to be used in a rocky field, another electrode may be more effective.

Bucket teeth, chisels, hammers, scraper blades, and other tools and parts subject to shock usually take a moderate

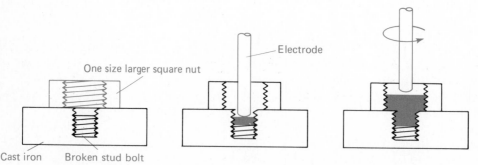

Figure 18-13. Removing a broken stud bolt with the arc welder should be used as a last alternative.

BASIC ARC WELDING

abrasion- and impact-resistant electrode.

The severe impact and moderate abrasion electrodes provide a tough and wear resistant surface. The surface resists cracking when struck. Knives and rasp bars are two examples where these electrodes may be used.

Proper preparation of the surface is very important in hard-facing. Use a grinder to remove all rust, grease, and oxides. Before facing, make a soapstone line to serve as a guide.

Use a long arc length when hard-facing. It is necessary to bond the alloy to the base metal without penetrating very deep. If the deposit penetrates too deeply, the alloy will mix with the base metal, and will result in a soft surface. Use a wide, semicircular motion to produce a flat bead. If necessary, a second bead may overlap the first one by about $1/8$ in [3.2 mm].

Some parts and tools, such as chisel shanks and scraper blades, have the alloy applied to their face. But in other cases, as with plowshares, sweeps, and shovels, the alloy is applied to the back side. This allows the front edge to wear faster than the back, thereby maintaining a sharp cutting edge.

Safety Rules

These safety rules should be followed in performing special arc welding jobs:

1. Wear protective clothing.
2. Be sure that the molten metal drops in a proper container when cutting with the arc.
3. Be sure there is proper ventilation. Much smoke is given off during the cutting process.
4. Do not use an arc welder to cut up barrels or other containers that have been filled with fuel or other flammable materials.

Out-of-Position Welding and Special Arc Welding Applications: A Review

Because of the size and complexity of agricultural equipment and machinery, not all welding can be done in the preferred flat position. It is necessary, therefore, to perform some welding in other positions. These positions include horizontal, vertical, and overhead.

The horizontal weld is made on a vertical surface, but the welded joint is in a horizontal plane. This weld joint is common in farming and agribusiness. The vertical weld, on the other hand, is both on a vertical surface and in a vertical plane. There are two types of vertical welds: vertical-up and vertical-down. The vertical-up weld produces more penetration but is more difficult to perform. The vertical-down weld can be performed faster than many other types of welds.

Pipe is used in the construction of agricultural machinery. When it is necessary to weld the pipe, all four positions may be needed: flat, horizontal, vertical, and overhead.

Special safety rules apply for overhead and pipe welding. When welding overhead, the worker should be careful to protect the top of the head and the arms.

The arc welder can be used to cut and pierce metal. It is particularly handy when the job must be performed on site instead of in the shop. However, the arc welder does not produce the same quality work as either the oxy-fuel torch or the saw.

The arc welder can be used to weld a square nut to the top of the stud when all other efforts have failed to remove a broken-off stud bolt.

The life of many tillage tools can be prolonged if they are hard-faced by using the arc welder. Blades, for example, stay sharper longer if they have been hard-faced.

THINKING IT THROUGH

1. Sketch a front and side view showing the proper position of the electrode in relation to the base metal for the following welds:
   (a) Horizontal butt weld
   (b) Horizontal lap weld
   (c) Vertical-down weld
   (d) Vertical-up weld
   (e) Overhead weld
2. What is the primary problem encountered in welding out of position?
3. What type of electrode should be used in welding out of position?
4. Using a pencil as an electrode, demonstrate how to perform a pipe weld.
5. List four examples of when you might be required to do out-of-position welding.
6. Describe the safety rules that you should follow when welding out of position.
7. Describe a situation when the arc welder, rather than a metal-cutting saw or drill, would be used to cut or pierce metal.
8. List the procedures for removing a broken stud bolt by using the arc welder.
9. What are some common applications for hard-facing in agricultural industry?
10. What safety precautions must be taken when containers such as barrels or cylinders are being cut?
11. What precautions should be taken in cutting or piercing with the arc welder?

UNIT VI

GAS WELDING, CUTTING, AND HEATING

COMPETENCIES

	PRODUCTION AGRICULTURE — Cash grain farmer	AGRICULTURAL SUPPLIES/SERVICES — Grain elevator operator	AGRICULTURAL MECHANICS — Tractor mechanic's helper	AGRICULTURAL PRODUCTS, PROCESSING, AND MARKETING — Processing plant maintenance mechanic	HORTICULTURE — Tree surgeon's helper	FORESTRY — Logger	RENEWABLE NATURAL RESOURCES — Wildlife conservation technician
Demonstrate safety when using the oxy-fuel torch	Very Important	Very Important	Very Important	Very Important	Important	Very Important	Important
Weld metal parts using the oxy-fuel torch	Very Important	Very Important	Very Important	Not Important	Important	Not Important	Not Important
Solder and braze sheet metal	Very Important	Very Important	Very Important	Not Important	Important	Not Important	Not Important
Demonstrate personal safety using the cutting torch	Very Important	Very Important	Very Important	Very Important	Very Important	Very Important	Important
Use the oxy-fuel torch for cutting metal	Very Important	Very Important	Very Important	Very Important	Important	Very Important	Important
Use the oxy-fuel torch for heating metal	Very Important	Very Important	Very Important	Important	Important	Very Important	Important

 Very Important
 Important
 Not Important

GAS WELDING, CUTTING, AND HEATING

Gas welding equipment has an advantage over arc welding equipment on some occasions, mainly because gas welding equipment is more portable. It is easier to transport gas cylinders than a portable electric welder. Also, a gas welder does a better job of heating and cutting.

As in the case of arc welding equipment, the gas welding or cutting torch has many uses in agricultural industry. For example, a farm worker may spend the off-season months in a farm shop repairing and remodeling equipment with the gas torch. Of course, during the months when the worker is outdoors, the portable gas equipment is very helpful when equipment breaks down in the field.

A grain elevator operator will use gas welding equipment to install such grain-unloading equipment as augers and conveyors. A logger may need to make on-the-spot repairs to a skidder in the forest. Instead of bringing in a generator for arc welding equipment, the logger may find it much easier to use portable gas welding equipment.

In general, the processes described in this unit are used frequently to construct, rebuild, and repair equipment, parts, machinery, and metal structures.

After you have mastered the skills outlined in this unit, you should know how to use an oxy-fuel torch (gas torch) to fusion-weld, cut, and heat metal. In addition, you should know how to join metals using brazing and soldering techniques.

After completing the unit, you also should know how to identify the correct equipment needed for specific jobs and be prepared to spot problems that can create weld failures.

Finally, you should know how to perform all operations safely.

As with any new techniques you are learning, it is helpful to observe others at work. You can gain much from watching agricultural workers use gas welding equipment.

CHAPTER 19

GAS WELDING EQUIPMENT

Gas welding techniques have been developed and improved quite rapidly during this century. Welding, however, was used in a crude way by blacksmiths in the early 1800s to supplement forge welding (commonly called just *forging*) In *forge welding,* a blacksmith simply heated the pieces of metal in a fire until they were almost white-hot. Then, using a hammer and resting the hot metal on an anvil, the blacksmith hammered the pieces of metal together.

Gases were sometimes used to create a hotter fire, just as they are today. Only the techniques and equipment have changed. The blacksmith used a carbide generator to produce a fuel gas. By dipping calcium carbide in water, the blacksmith could produce an acetylene gas. Because of improper use, many of these generators exploded, causing considerable injury and damage.

Every blacksmith understood and respected the carbide generator. The same respect should be given the modern oxyacetylene welding torch. When properly handled, it is a safe piece of equipment. Improperly handled, one acetylene cylinder has the same potential as five cases of dynamite.

Welders today have more control of the welding process because of the equipment that is available to them. These changes in technology have enabled the agricultural worker to use better equipment and do a much better job than the blacksmith of the 1800s.

The modern welder has more control over temperature. And the purity of the gases used is far superior to the gases produced by the early day suppliers. Also, the equipment does a much better job of regulating gas pressure and flame characteristics.

CHAPTER GOALS

In this chapter your goals are:

- To identify the basic gas welding equipment commonly used in agricultural industry
- To set up an equipment and check for gas leaks
- To properly light, adjust, and turn off the gas welding equipment
- To demonstrate the safety rules followed in using gas welding equipment

Gas Welding Process

All welding requires a high temperature heat source. In gas welding, burning gases create heat. Most agricultural welding jobs

GAS WELDING, CUTTING, AND HEATING

use a fuel called *acetylene* for welding. When combined with pure oxygen, acetylene burns with a very hot flame. When the two gases oxygen and acetylene are combined, they produce an oxyacetylene flame. The oxygen itself does not burn. It supports or makes possible the burning of other gases. Other cutting or heating jobs may use acetylene, Mapp-gas, liquid petroleum gas, or other synthetic gases as a fuel. These gases generally do not produce enough heat for welding jobs.

A neutral or balanced oxyacetylene flame burns at an average temperature of 5589°F [3087°C]. To get a rough idea of how hot that is, consider that water boils when the temperature reaches only 212°F [100°C]. The oxyacetylene flame is hot enough to melt most of the metals ordinarily used in agricultural equipment and machinery.

Many Possible Uses

Because of its high temperatures, gas welding is a versatile process. That means it can be used for cutting steel and welding most kinds of metal. Furthermore, it has the advantage of being more portable than arc welding equipment.

Gas welding equipment can easily be taken out into the field to work on broken machinery. For that reason, it is used by the agricultural worker and mechanic.

Cylinders, Regulators, Hoses, and Torch

Gas welding equipment includes cylinders, regulators, hoses, and a torch. The torch, of course, is that piece of the equipment through which the gases are passed to do the work. All parts, however, have to work properly if the equipment is to do what is expected.

Cylinders

The cylinder is the container used to hold the gas, and it comes in many sizes and colors. One can get cylinders with small volumes of gas for small jobs or for jobs that require the cylinders to be moved. Larger cylinders can be obtained that contain greater volumes of gas. The larger cylinders are used when there is a lot of work to be done. In some welding shops where there are several welding stations, a *manifold system* is used. This means simply that several cylinders are collected at a central spot. They are connected, and the gases are piped to the different welding stations. These are often referred to as a *fixed manifold system*. The system eliminates the need to have several cylinders in the shop at one time.

Oxygen Cylinders. Cylinders do come in all sizes, but as a rule oxygen cylinders are usually taller and narrower than cylinders holding the fuel gases (see Figure 19-1). However, one cannot rely on shape or color to tell what gas is in the cylinder. Because all gases are potentially very dangerous, extreme care must be taken when working with them. Always know exactly what gas is in the cylinder.

Cylinders are always labeled, describ-

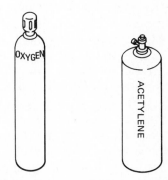

Figure 19-1. Oxygen cylinders are taller and narrower than cylinders holding other kinds of gases.

ing the gas contained, but there are other identifying characteristics to prevent a mistake. Industrial oxygen cylinders are usually painted green.

A valve with a hand wheel sits on top of the oxygen cylinder. The valve has an outlet on one side where the regulator (described later) is connected. There is a safety, or "pop-off," outlet on the other side of the valve. The safety outlet is extremely important, because the oxygen is kept under very high pressure in the cylinder. If the cylinder were heated accidentally, the pressure would increase. If that happens, the safety outlet blows open and allows the oxygen to escape harmlessly. Without this safety outlet, the cylinder could explode if the pressure were increased significantly. The valve of the oxygen cylinder always has right-hand connector threads.

The oxygen cylinder is designed to hold the gas at approximately 2200 psi [152 bars] at normal room temperature (about 70°F, or 21°C). How much pressure is this? Well, if the valve at the top of a full cylinder of oxygen were accidentally broken off, the force of the escaping gas would be strong enough to launch a fully charged cylinder like a rocket, with speeds of over 200 miles per hour (mph) in the first second.

To prevent dangerous accidents like this, a federal law requires oxygen cylinders to have a protective cap that screws on over the valve. When the cap is removed and the cylinder is being used, it must be chained in an upright position to prevent it from falling over.

Always open cylinder valves slowly, and never stand in front of the outlet when opening a valve. Before using the cylinder, always *crack* it first. This means opening the valve quickly to blow out any dirt. Never stand in front of the outlet when cracking a valve. The valve should be fully opened when the cylinder is in use. The oxygen valve seats at both the open and closed position to prevent the loss of oxygen around the valve stem.

Oxygen cylinders are made of steel. The most common cylinder size is 244 cubic feet (ft³) [6.91 m³]. Other sizes are 10, 122, and 330 ft³ [0.28, 3.46, and 9.35 m³].

Acetylene Cylinders. Acetylene cylinders generally are shorter and wider than oxygen cylinders and are usually red. The cylinder will have the word *acetylene* stenciled on the body or will have a tag properly labeled. When in use, acetylene cylinders are chained upright with the oxygen cylinder.

Acetylene is not kept under as great a pressure as oxygen. Acetylene cylinders are filled to about 250 psi [17.2 bars]. The most common acetylene cylinder size has a capacity of about 125 ft³ [3.54 m³], although, as with oxygen cylinders, there are different sizes. Many suppliers sell acetylene by weight rather than volume.

Because acetylene is an *unstable* gas—one that decomposes easily at increased pressure—it cannot be used above a pressure of 15 psi [1.04 bars]. At pressures above this point, the gas is likely to explode. To prevent this and to keep the acetylene below this pressure, the acetylene cylinder is filled with a *porous* material—one that has a lot of holes in it much like a sponge. The cylinder is filled with a liquid called *acetone*. This liquid absorbs the acetylene gas and holds it until it is withdrawn. Acetone will absorb 15 times its own volume of acetylene. The acetylene withdrawal rate should not exceed one-seventh of the cylinder capacity.

It is because of the liquid acetone that one must always keep an acetylene cylinder upright. If the cylinder is turned on its side, the acetone can flow into the cylinder valve and through the welding torch. That would cause damage to the regulator and hoses. If the acetylene cylinder has been

GAS WELDING, CUTTING, AND HEATING

lying down for any reason, it should be allowed to stand upright at least 30 minutes at room temperature (about 70°F, or 21°C) before using it.

Open the valve of the acetylene cylinder slowly. Never open it more than one turn. Before using, crack it as you would the oxygen cylinder to clean out any loose foreign material and to check for excess acetone. If acetone is spit from the valve, return the cylinder to the gas supplier. Be sure no one stands in front of the outlet when the valve is cracked.

Other fuel gases such as MAPP and propane are available in liquid petroleum gas (LPG)-type cylinders and are sold on the basis of weight. Because these cylinders are not filled with acetone, they are much lighter, easier to handle, and cheaper to transport. The most common size is the 70-lb [31.8-kg] cylinder, but sizes range from 5 to 200 lb [2.3 to 90.7 kg].

Other fuel gases use different color designations. MAPP gas is always yellow, but other fuels may be available in blue or orange cylinders.

Regulators

The name *regulator* describes what this device does: it regulates the pressure and flow of gas from the cylinder. Underwater divers use a regulator to adjust the flow of air from the tanks on their backs. An oxygen or acetylene regulator serves the same purpose. The cylinder pressures are too high to be used directly. The regulators can be adjusted, therefore, to reduce the pressure.

Most regulators have two *gauges* on them, although the regulator used on a manifold system has only one gauge. Gauges are simply measuring devices. They have a numbered face and a needle that points to the number to give a reading. Since these gauges measure pressure, the numbers refer to pounds of pressure per square inch (psi) [or bars, grams per square centimeter (kg/cm^2)].

One gauge measures the pressure of the gas in the cylinder. As gas is used, the pressure drops, so this gauge tells how much pressure is left. This gauge is not

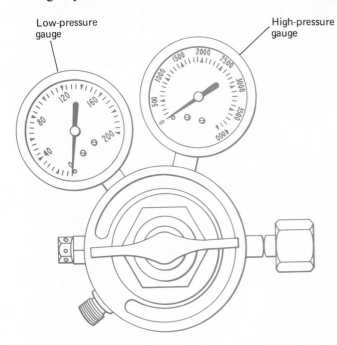

Figure 19-2. An oxygen regulator reduces the 2200 psi cylinder pressure for welding, cutting, or heating jobs.

necessary on the manifold system since all welding stations use the same cylinders. The oxygen gauge is calibrated to 4000 psi [276 bars], which means it is designed to measure up to 4000 psi of pressure. Each gradation usually represents 100 psi, [6.9 bars].

The second gauge measures the pressure of the gas that is available to the welding torch. This gauge (shown in Figure 19-2) is usually calibrated to 100 psi and each gradation represents 5 psi [0.35 bars]. The adjusting screw on the regulator works against spring tension that allows one to increase or decrease pressure as needed. When the screw is turned clockwise and spring tension is increased, the gas pressure to the torch also is increased. When the screw is turned counterclockwise,

there is no pressure against the spring and no gas pressure. To make a good weld, the gases must be mixed and at the right pressure when delivered to the torch.

Two Regulators. There are two kinds of regulators: *single-stage* and *two-stage*. They differ in how they work to reduce or regulate the flow of pressure to the welding torch. The single-stage regulator (shown in Figure 19-3) reduces the pressure in one step. The pressure-adjusting screw controls the single valve opening that allows the gas to go out of the cylinder and to the torch. In the two-stage regulator, the gas is released in two stages. First, a preset valve lets it into a chamber where the pressure is always constant, regardless of how much gas has been used from the cylinder. The

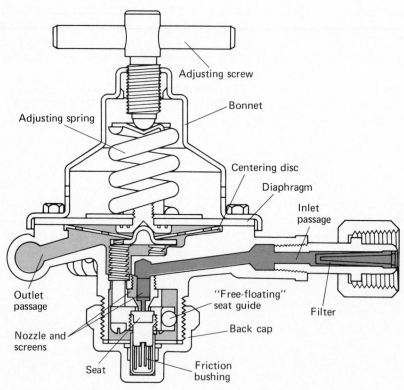

Figure 19-3. A cut-a-way view of the single-stage regulator.

GAS WELDING, CUTTING, AND HEATING

pressure-adjusting screw lets the gas out of that chamber to the torch at the pressure desired.

The single-stage regulator is less expensive. The two-stage regulator helps to constantly maintain the right pressure for welding, even as the amount of gas in the cylinder decreases.

Just as oxygen cylinders are usually painted green, so are oxygen regulators. All are threaded right-hand to match the right-hand threads of the oxygen cylinder. And they have a female inlet with threads on the inside of the connector nut that match the male threads on the outside of the oxygen cylinder.

Acetylene regulators are usually painted red. They have left-hand threads and usually have a male-inlet connecting nut. As a further safety precaution, the brass nut that joins the regulator to the cylinder has a V-notch cut around it to identify it as a left-hand threaded regulator for fuel gases.

As all these precautions indicate, it is very important not to confuse an oxygen regulator with a fuel gas regulator. The different pressures under which the different gases are kept and used make it necessary for each gas to have its own regulator.

All regulators are delicate, precision instruments. Handle them with care. The body of the regulators and their connections are made of brass. One reason is that brass will not give off a spark if accidentally hit with another piece of metal, such as a wrench. Obviously, even a tiny spark

could be very dangerous when close to such highly combustible gases.

Never oil the adjusting screw or connections. But keep them free of dust and dirt. If a regulator does not work as it should, do not try to fix it. Contact an authorized service center or representative who has special training.

Hoses

Hoses carry the gas from the cylinders to the welding torch. The flexible rubber hoses allow the welder to move around and weld in different locations. The long hoses also help keep the hot welding flame safely away from the gas cylinders. Usually these hoses are about 25 ft [7.63 m] long. When not in use, the hoses are neatly coiled and hung on the hose rack of the cylinder cart.

To continue the system of color identification, the oxygen hose is green and has right-hand threads. The fuel gas hose is red and has left-hand threads. The two hoses are molded or clamped together for easier handling.

Use only approved hoses, which are flame-retardant. That means that, although they will burn if a flame is turned on them, they will not continue to burn if the heat source is taken away. Always be careful not to accidentally direct the welding torch on the hose.

Protect the hose from molten metal, sparks, and open flames. If it does become damaged, do not try to patch it. Replace it. To splice, or join, two good pieces of hose,

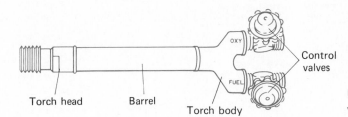

Figure 19-4. A Victor welding torch handle.

Torch head   Barrel   Torch body

Control valves

use only an approved, barb-type hose splicer. Hoses that have many splices are not satisfactory and should be replaced.

New hoses have talcum powder in them. This must be blown out before the welding torch is attached. Always keep hoses away from oil and grease that may be on gloves or the floor. These agents can destroy a rubber hose and may burn readily in the presence of oxygen.

Welding Torch Handle

The welding torch handle (see Figure 19-4) is what is used to actually make the weld. There is a Y at the gas inlet end. Each arm of the Y has a control valve. One arm is marked "Oxy," and it accepts right-hand threads. The other arm is generally marked "Fuel," and it has left-hand threads. This is where the hoses attach. The valves on the Y arms control the flow of oxygen and fuel gas to get the right combination. These gases are not mixed in the torch body, or *handle*, as it is sometimes called. A small metal tube carries the oxygen through the handle to keep it separate from the fuel gas (see Figure 19-7).

A small, metal element that looks like an extension of the body, called the *torch head*, may be threaded onto the end opposite the Y or it may be built right into the body. The oxygen passes through a center hole in the head, while the fuel gas passes through holes drilled around this center hole. The welding tip, or blow pipe, screws into the torch head.

Welding Tips and Cutting Attachments

Depending on the job the worker is doing, either a welding tip or a cutting attachment will be used with the torch body.

Welding Tips

The unit shown in Figure 19-5 goes on the end of the torch handle and is called the *welding tip*. The tip is a soft copper pipe that carries the gases to the orifice (opening). The gases are ignited as they blow out from the tip (causing the tip sometimes to be called the *blow pipe*). A *friction lighter*, also called a *spark lighter*, is used to light the fuel. Never use a match, open flame, or cigarette lighter because you could be severly burned (see Figure 19-6).

The size of the hole, or orifice, in the tip varies. Some tips have very small openings, and others have relatively large openings. Naturally, the size of the opening in the tip determines the size of the flame. The temperature of the flame remains the same regardless of its size. Only the volume of heat—measured in British thermal units (Btu's) or megajoules—changes, so a

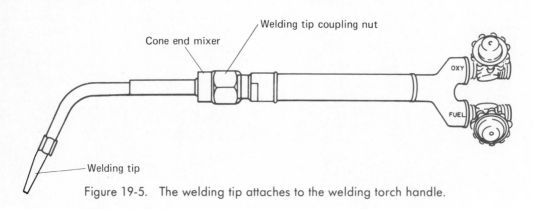

Figure 19-5. The welding tip attaches to the welding torch handle.

GAS WELDING, CUTTING, AND HEATING

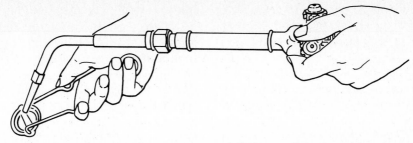

Figure 19-6. Always use a spark lighter to light the torch.

larger tip only brings the metal joint to the welding temperature faster. Tip size depends on the thickness of the metal being welded.

The hole in the welding tip sometimes becomes blocked. The hole must be kept open and clean. To clean holes in welding tips, use a wire brush and rasp-type tip cleaner. Be careful to select a tip cleaner that is the right size for the tip being cleaned. It is possible to misuse the cleaner and enlarge the hole or break off the tip cleaner in the orifice.

Cutting Attachments

In cutting a piece of metal, a special cutting torch or cutting attachment is added to the regular torch handle. When a welder cuts a piece of metal, the process may appear to be similar to welding. It seems that the welder is simply applying a hot flame to the metal and letting it cut through. But there is a big difference in the flame. Pure oxygen is being metered through the flame to the metal. This oxygen jet actually completely burns the heated metal. In welding, no pure oxygen reaches the metal. If it did, it would produce a weak, rather than a strong, weld.

A special cutting attachment like the one shown in Figure 19-7 delivers a stream of oxygen along with the preheating flame to the metal being cut. Cutting attachments have one extra control valve. Two valves control the flow of oxygen, and one con-

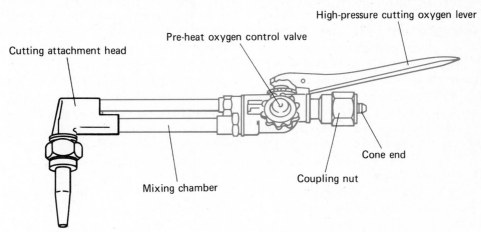

Figure 19-7. A cutting attachment is added to the welding torch handle to make the cutting torch.

trols the flow of fuel gas. In addition, the cutting attachment has a control lever which is pressed to open the jet of oxygen in cutting.

Cutting heads generally are at a 90° angle to the torch body, although one may see some that are at 75° or 180° angles to the torch body.

Always inspect the cutting attachment for damaged *seats* (the parts that fit on to the rest of the torch) or damaged O rings (sealing rings made of rubber or neoprene) to make sure that connections will allow no gas to escape. Any damaged part should be repaired or replaced. A poor fit can cause improper mixing of the gases and a backflash.

A *flashback* makes a hissing sound. It is caused by the flame burning inside the nozzle or tip. This is often caused by an overheated tip. Overheating usually results from an insufficient volume of gas being available to the torch. When a backflash occurs, turn off first the fuel, then the oxygen. Allow the torch to cool. Then inspect the complete torch for dirty or damaged parts, and for correct working pressure settings.

Procedures for Setting Up the Torch

Check and follow all safety rules for using the oxy-fuel welding torch.

Before attaching the regulators to the cylinders or manifold, crack the cylinder valves to blow out any dirt or foreign material. Always stand to one side of the valve. Attach the proper regulator to each cylinder. Use the correct-size wrench to tighten the inlet connections. Make sure connections are snug but not too tight. Attach the hoses to the regulators—the green hose on the oxygen cylinder and the red hose on the fuel cylinder.

Blow talcum powder out of the new hoses before attaching the welding torch body. Attach the torch body. The green hose goes on the oxygen control valve marked "Oxy." The red hose goes on the fuel control valve marked "Fuel." Select and attach the proper-size welding tip for the metal. Tighten the welding tip according to the manufacturer's recommendations. Some are recommended to be tightened by hand only.

Make certain the spring tension on both pressure-regulating screws on the regulators is released by turning the screw counterclockwise until all pressure is released. In opening a cylinder against a preset regulator, over 150 tons per square foot (tons/ft²) [143.75 bars] of pressure is exerted against the diaphragm. This may cause the regulator to explode. Standing with the cylinder valve between you and the regulator, open the oxygen cylinder valve very slowly until maximum pressure is registered on the high pressure gauge. Now open the cylinder valve fully to seat the high pressure valve.

Check the fuel gas cylinder pressure-regulating screw valve in the same way. Acetylene cylinder valves should be opened only one turn. Other fuel gas cylinder valves such as MAPP or propane should be opened fully.

Check for leaks around the inlet connections using a small brush and an approved leak-detector solution. Many manufacturers recommend a solution of 1 part Ivory soap®, which does not contain oils, with 4 to 6 parts water. Brush the solution over all connections. Any leaks will cause bubbles. Regulator connections may be tightened if leaks appear. If the cylinder valve shows a leak, set the cylinder outside and tell the instructor. *Do not attempt to tighten a cylinder valve.*

Adjust the oxygen pressure-regulating screw to allow 5 psi [0.35 bars] to escape through the hose. Using the solution, check all connections for leaks. Close the oxygen control valve. Close the fuel control valve. Adjust the fuel pressure-regulating screw to 5 psi, and check all connections for leaks.

Lighting the Torch

The following steps should be followed when you light the torch:

Open the oxygen control valve and adjust the regulator screw to obtain the working pressure, and make final adjustments. Close the oxygen control valve and open the fuel regulator valve and the fuel control valve; make final adjustments for working pressures. Close the fuel control valve.

Be sure to wear welding goggles to shield the eyes from the bright flame. Hold the torch in one hand and the spark lighter in the other. Open the fuel control valve about one-half turn and ignite the gas, pointing the flame away from the cylinders. Quickly open the fuel control valve until the flame is clean-burning.

Open the oxygen control valve until a bright, neutral flame is reached.

Turning Off the Torch

The following steps should be taken to turn off the torch:

Close the fuel control valve, extinguishing the flame, and close the oxygen control valve. Close the fuel cylinder valve and open the fuel control valve, allowing the fuel gas to escape until both pressure gauges on the fuel regulator read zero. Close the fuel regulator by turning the pressure-regulating screw counterclockwise until it is free.

Close the fuel control valve and open the oxygen control valve, allowing the gas to escape until both pressure gauges read zero. Close the oxygen regulator by turning the pressure-regulating screw coun-

terclockwise. Close the oxygen control valve. Hang the hose and the torch in their proper place.

Safety Rules

The following safety rules should be followed in using gas welding equipment:

1. Always use safety goggles when working in the shop. When you are using the welding torch, a shade number 5 or darker lens should be worn.
2. Wear leather gloves to protect the hands from hot metal.
3. Long sleeves, cuffless pants, and leather boots should be worn to protect the body from hot metal.
4. Clothing should be fire-retardant. Many synthetic materials have a low melting point and, if overheated, can stick to the skin and cause a bad burn.
5. Clothing should be free of grease and oil, since they can be ignited.
6. Do not use oxygen to blow dirt off clothing or to cool the body.

Gas Welding Equipment: A Review

The most common gases used in gas welding and cutting are oxygen and acetylene. Several new fuel gases are available which may be substituted for acetylene on cutting or heating jobs. All gases are stored in heavy steel cylinders. These cylinders are under very high pressure and require special care in handling.

Because of the high pressure, both oxygen and the fuel gas require special regulators to lower the working pressure. The oxygen regulator is usually green and has right-hand threads to attach to the cylinder. The acetylene regulator is usually red and has left-hand threads. Other special regulators are available for the other fuel gases, but they vary in both color and type of thread. Always check the manufacturer's recommendations concerning regulator care.

The gases are moved to the torch through special hoses which are red and green and usually molded into one piece. These hoses should be protected from open flames and grease. The hoses attach to a welding torch body which is used for both welding and cutting. Care should be taken to be sure the green hose is attached to the oxygen inlet and the red hose is attached to the fuel inlet.

Welding tips are selected according to the volume of heat which is required—not the temperature. The temperature of the flame depends on the type of fuel gas that is used, while the volume of heat depends on the size of the tip. Always consult a chart to determine the correct tip size and gas pressures for a specific job. A special attachment is used for cutting steel that connects directly to the welding torch handle. This attachment has an extra control valve to regulate the flow of oxygen to the preheating flame of the cutting tip.

There are specific rules which should be followed in setting up equipment, lighting the torch, and turning off the equipment.

Learn and practice the safety rules when you are using gas welding equipment.

THINKING IT THROUGH

1. Discuss why acetylene cylinders should be handled with special care.
2. Describe the advantages and disadvantages of a single-stage versus a two-stage regulator.
3. What are the differences between an acetylene cylinder and a MAPP-gas cylinder?
4. What are the differences between a welding torch and a cutting torch?
5. What procedures should be followed in setting up the oxyacetylene welding torch?
6. What procedures should be used to turn off the welding outfit?
7. List the safety precautions which should be observed in using gas welding equipment.
8. What gases can be used for cutting steel?

GAS WELDING, CUTTING, AND HEATING

CHAPTER 20

FUSION WELDING

The fact that gas welding equipment is more portable is one of its advantages over arc welding equipment. Another advantage is that the oxyacetylene torch produces a weld that is often stronger than the original metals. A disadvantage of gas welding compared to arc welding is that it takes much longer to make a weld.

The oxyacetylene torch generally is used to weld 7-gauge sheet steel (less than $^3/_{16}$ in, or 4.8 mm, thick). However, the torch also can be used to weld pieces of steel plate (metal thicker than $^3/_{16}$ in) and to join aluminum or other similar metals.

Here is just one situation in which an agricultural worker could use an oxyacetylene welding torch: a field of hay is being mowed when suddenly the mower hits a hidden rock that breaks the outer shoe. The worker decides the shoe can be repaired in the field by welding. The worker gets the portable gas welding equipment and brings it to the field, where the welding is done. After a minimum of lost time, the mower is ready to use again.

CHAPTER GOALS

In this chapter your goals are:

- To prepare metal for welding and select the correct pressures, tip, and welding flame for a specific job

- To know how to make any of four weld joints in any of the three common welding positions
- To diagnose common welding problems and prescribe the best solution to correct the problems

Preparing the Metal

The pieces of metal to be fusion-welded are called the *base metal*. A good weld cannot be made unless the base metal is "shiny" clean. To prepare the base metal, first remove all mill scale, rust, dirt, paint, and grease using a power grinder or wire brush.

If the base metal is $^1/_4$ in [6.4 mm] or thicker, it will require beveling. *Beveling*, as explained in Unit V and Chapter 22, is cutting or grinding the edge of the metal at an angle. When the base metal is more than $^1/_4$ in thick, it is necessary to make a 60° bevel cut along the edges to be joined to allow the filler metal to penetrate the two pieces of base metal (see Figure 20-1). Normally the root opening—the space between the two pieces of metal—should be about one-half the thickness of the base metal or the thickness of the shoulder of the bevel cut.

If the base metal is over $^1/_2$ in [12.7 mm] thick and is to be welded from both sides, the 60° bevel cut should be made on both sides (see Figure 20-1).

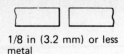

1/8 in (3.2 mm) or less metal

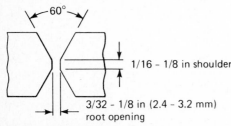

1/4 in (6.4 mm) metal

1/16 – 1/8 in (1.6 – 3.2 mm) shoulder

1/16 – 1/8 in shoulder

3/32 – 1/8 in (2.4 – 3.2 mm) root opening

1/2 in (12.7 mm) metal

Figure 20-1. Procedures used when preparing the base metal for fusion welding.

TABLE 20-1. Cutting Orifice Reference Chart
(For all standard cutting tips except machine torch high-speed models)

CUTTING TIP SIZE	ORIFICE DIAMETER (IN)	ORIFICE SIZE (ACTUAL)
000	.027	·
00	.033	·
0	.041	·
1	.047	·
2	.060	·
3	.071	•
4	.083	•
5	.100	●
6	.121	●
7	.141	●
8	.162	●
9	.173	●
10	.185	●
12	.221	●

Source: Victor Welding Co.

Selecting the Correct Tip and Pressure

The size of the welding tip regulates the amount of heat available for the welding process. The larger the tip, the more heat is available. Table 20-1 shows different tip sizes. Tip sizes are designated by numbers, with the lowest number having the smallest opening (orifice). Because designations vary among manufacturers, always consult the specific torch recommendations. As noted in the table, a number 1 tip has a orifice diameter of 0.047 in [1.2 mm].

Sometimes the tip size is designated by the drill size of the opening. For example, a drill size of 65 would be equivalent to a number 0 tip. Tips commonly used in agricultural welding jobs are numbers 1, 3,

The last step in the preparation of the base metal is to tack-weld every 12 to 14 in [30.5 to 35.6 cm] along the weld joint to hold the base metal together.

and 5. Some special jobs require larger tip sizes; these will be discussed later in the unit.

The thickness of the base metal is the primary factor in selecting the welding tip. The thicker the base metal, the greater the volume of heat required. The more heat required, the larger the tip. In welding 1/8-in [3.2 mm] mild steel, for example, a number 3 tip would be required. Table 20-2 can be used as a guide in tip selection.

Sometimes the tip opening becomes clogged with molten metal or dirt. When this happens, the opening should be carefully cleaned with the proper-size wire tip cleaner (see discussion of wire tip cleaner in Chapter 19). If the tip is very badly clogged, it may be necessary to use a drill to clear the obstruction. If a drill is used, select one that matches the size of the opening. Table 20-1 gives the correct drill sizes for the various tip sizes.

After the tip size has been selected, it is important to use the correct gas pressures for the welding job. The pressures depend largely on the thickness of the base

GAS WELDING, CUTTING, AND HEATING

TABLE 20-2.   Acetylene Welding Nozzles

METAL THICKNESS	TIP SIZE	DRILL SIZE	ACETYLENE PRESSURE (psig) MIN./MAX.
Up to $1/32''$	000	75	$3/5$
$1/64''$-$3/64''$	00	70	$3/5$
$1/32''$-$5/64''$	0	65	$3/5$
$3/64''$-$3/32''$	1	60	$3/5$
$1/16''$-$1/8''$	2	56	$2/5$
$1/8''$-$3/16''$	3	53	$3/6$
$3/16''$-$1/4''$	4	49	$4/7$
$1/4''$-$1/2''$	5	43	$5/8$
$1/2''$-$3/4''$	6	36	$6/9$
$3/4''$-$1 1/4''$	7	30	$7/10$

Most oxyacetylene welding torches are equal pressure designs. Always set the acetylene and oxygen regulators at the same pressure.

metal and the welding tip size. If acetylene is used at a pressure of 15 psi [1.04 bars] or more, it becomes unstable and an explosion is possible. The pressures listed in Table 20-2 are recommended working pressures. The final working pressures are set by adjusting the regulator after the torch has been lit.

Selecting the Flame

Three types of welding flames can be used for agricultural jobs: *neutral*, *oxidizing*, and *carburizing*. The ratio of oxygen to the acetylene gas determines the type flame.

Neutral Flame

The *neutral flame* (see Figure 20-2) is the one most commonly used in welding. It is the most effective in fusion-welding mild steel. The neutral flame consists of equal parts of oxygen and fuel gas. The inner cone of the flame (that portion closest to the tip of the torch) has a blunt end and a temperature of approximately 5600°F [3093°C]. The flame envelope—the larger part of the flame—is generally light blue. The neutral flame consumes all the oxygen

in the air around the weld area. This prevents oxygen from combining with the molten metal, which would weaken the welded joint.

Oxidizing Flame

The *oxidizing flame* (see Figure 20-3) is rarely used in fusion welding. Its main use is limited to braze welding, which is discussed in the following chaper. The oxidizing flame also has no feather, but the inner cone is pointed rather than blunt. The flame envelope is a very light blue, sometimes appearing almost white. The oxidizing flame is the hottest of the three flames, the temperature reaching about 6000°F [3316°C].

Carburizing Flame

The *carburizing flame* (see Figure 20-4) is sometimes called a *reducing flame*. It is

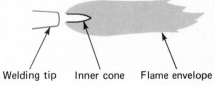

Welding tip   Inner cone   Flame envelope

Figure 20-2.   A neutral oxyacetylene flame.

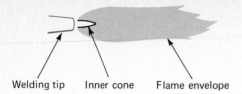

Welding tip   Inner cone   Flame envelope

Figure 20-3. An oxidizing flame has a sharp point.

used mostly to heat metal, soft-solder, and hard-surface. The carburizing flame has the blunt inner cone, but there is also a feather indicating an excess of acetylene. This excess acetylene deposits carbon on the surface layer of the metal. Because this layer melts at a lower temperature, the penetration of the weld metal is usually shallow. The flame envelope is orange. Carburizing flames often are designated 1X, 2X, or 3X. The numbers refer to the ratio of the length of the acetylene feather to the length of the inner cone. This is an indication of the oxygen-reducing effect of the flame. For example, a 2X flame has a feather that is 2 times the length of the inner cone.

Selecting the Welding Rod

Always use a rod of a metal similar to the base metal. For example, in welding mild steel, use a steel rod. The thickness of the base metal determines the size of the welding rod. If the base metal is 16 gauge or thinner, use a $1/16$ in [1.6 mm] rod. If the

base metal is between $1/8$ and $1/4$ in [3.2 and 6.4 mm] thick, the rod should be $1/8$ in. If the base metal is thicker than $1/4$ in, select a rod $5/32$ in [4.0 mm].

It is a good safety practice to bend the upper end of the rod into a hook to prevent it from gouging the welder or someone else in the eye.

Welding Techniques

There are two gas welding techniques: *forehand* and *backhand*. The technique selected depends largely on the thickness of the base metal. The forehand technique is generally used when the base metal is $3/16$ in [4.8 mm] thick or less. The backhand technique is used when the base metal is thicker than $3/16$ in.

Forehand Technique

When the forehand technique is used, the flame is pointed in the direction in which welding will progress (see Figure 20-5). The flame is moved across the joint to pre-heat the edges. The filler rod is held just ahead of the flame. Before you begin to weld, the base metal should be tacked on the left end of the weld. Then, start welding on the right end with the tip pointed toward the leading edge of the molten puddle.

Backhand Technique

The backhand technique (see Figure 20-6) is used when the base metal is more than $1/8$

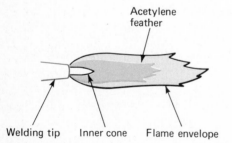

Welding tip   Inner cone   Flame envelope

Figure 20-4. A carburizing flame has excess acetylene.

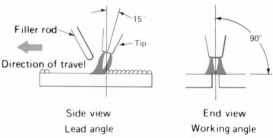

Figure 20-5. The forehand technique.

GAS WELDING, CUTTING, AND HEATING

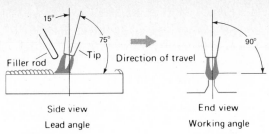

Side view

Lead angle

End view

Working angle

Figure 20-6. The backhand technique.

in [3.2 mm] thick. Tack the base metal at the right end of the weld. Then begin at the left end with the tip pointed toward the trailing edge of the molten puddle.

Butt Welds

A *butt weld* is one that joins two pieces of metal on the same plane so that the edge of one piece is against the edge of the second piece. The butt weld may be done where the base metal is flat, horizontal, vertical, or overhead. It is easier, of course, to weld in the flat position. But this is not always possible, especially if the worker is using the welding torch to repair equipment in position.

Flat Butt Weld

A flat butt weld (see Figure 20-7) using a filler rod and the forehand technique is the most common form of gas welding for agricultural jobs. The welder controls the molten puddle according to (1) the amount

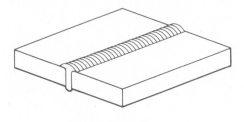

Flat butt weld

Figure 20-7. A flat butt weld is the most common form of gas welding.

of heat used, (2) the speed with which the weld bead is moved forward, (3) the distance between the inner cone of the flame and the molten puddle, and (4) the angles of the welding tip and filler rod.

The volume of heat available is directly related to the size of the torch tip. The larger tip opening will produce a greater amount of heat. The temperature of the inner cone remains constant and is not affected by the size of the tip. Only the volume of heat is affected. When acetylene is used, the temperature remains at approximately 5500–6000°F [3038–3316°C]. If the force of the flame tends to blow the molten puddle away, a larger welding tip should be used. If the tip makes a popping noise, there is not sufficient fuel available and it is necessary either to select a larger tip or to correctly adjust the working pressure of the one being used.

The speed with which the torch is moved is important. Incorrect speed will cause breaks in the ripple pattern of the weld. The puddle is formed with the welding rod at the initial point of the weld. Then the tip of the inner cone of the flame should be moved along the leading edge of the puddle. This technique will produce a butt weld with a uniform shape, width, and height.

The distance between the tip of the inner cone and the molten puddle determines the size of the puddle and, to some extent, the temperature of the base metal. The tip of the inner cone is the hottest part of the flame. It should never touch the base metal or molten puddle. If the molten puddle is too small, then the inner cone is too far from the puddle or the tip is too small.

Although most of the heat from the flame should be directed at the puddle, some is used to preheat the base metal ahead of the weld. For example, with the forehand technique, when base metal that is $^1/_8$ in [3.2 mm] thick is welded, the tip of the torch should be held at a 15° angle in the direction of travel. Such an angle will maintain the puddle properly and also pre-heat the base metal ahead of the weld.

Procedure for Welding. The flat butt weld on $^1/_8$ in [3.2 mm] metal is common on agricultural jobs. Here are the basic steps to follow for this kind of welding:

Using Table 20-2, select a number 3 tip. Set the oxygen and acetylene regulators at 5 psi [1.04 bars]. Check the torch manufacturer's recommendation for specific pressure settings. The base metal should be tacked at 12 to 14 in [30.5 to 35.6 cm] intervals. The root opening between the two pieces of metal should be approximately the thickness of the base metal. The $^1/_8$-in [3.2-mm] base metal is positioned so that the weld joint is at a right angle to the person welding.

If you are right-handed, start the weld at the right end of the joint, with the welding tip held at a 15° angle to the weld and at a right angle to the base metal. Holding the tip of the flame's inner cone approximately $^3/_{32}$ in [2.4 mm] from the edge of the base metal, move the tip back and forth across the joint until the base metal flows into a small puddle. Holding the end of a $^1/_{16}$-in [1.6-mm] mild-steel filler rod at a 45° angle in the flame envelope, dip the rod in and out of the puddle. The puddle will melt the end of the rod and add the metal to the bead of the weld.

Continue to advance the weld approximately $^1/_{16}$ in with every motion. Since the forehand technique preheats the base metal, the speed of travel should increase somewhat as one nears the end of the weld. Because the base metal is being preheated, it may be necessary occasionally to raise the inner cone slightly and increase the amount of filler rod to avoid burning through the base metal. Completely fill the molten puddle at the end of the butt weld to ensure that the weld is complete and will hold.

The backhand technique used for welding follows most of the same procedures, with some important exceptions. First, select a larger welding torch tip if you are welding thicker metal. For $^1/_4$-in-thick [6.4-mm] plate, use a number 4 tip. Set the oxygen and fuel pressures at 7 psi [1.50 bars] for base metal that is $^3/_{16}$ to $^1/_4$ in [4.8 to 6.4 mm] thick. Select a $^1/_8$ or $^5/_{32}$-in [3.2 or 4.0-mm] filler rod. Bevel the base metal as shown in Figure 20-1, leaving a root opening of approximately half the thickness of the metal.

The backhand technique should start on the left side of the joint if the welder is right-handed. Hold the welding tip at a 15° angle. As the tip moves away from the tack weld, increase the welding tip angle to approximately 30°. The filler rod should be

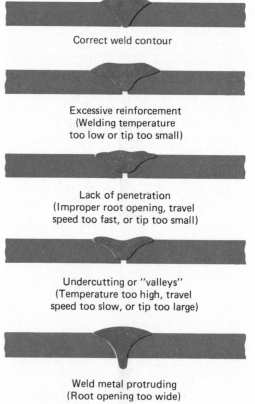

Correct weld contour

Excessive reinforcement
(Welding temperature
too low or tip too small)

Lack of penetration
(Improper root opening, travel
speed too fast, or tip too small)

Undercutting or "valleys"
(Temperature too high, travel
speed too slow, or tip too large)

Weld metal protruding
(Root opening too wide)

Figure 20-8. Common oxyacetylene welding problems.

GAS WELDING, CUTTING, AND HEATING

held near the trailing edge of the molten puddle.

When butt welding in the flat position is mastered, practice doing it with the base metal in a horizontal or vertical position. This is harder to do, of course, because the molten metal tends to run down. However, it's important for many agricultural workers to be able to weld in these positions. Many times it is impossible to get the joint in a flat position (see Figure 20-8.)

Corner Welds

Sometimes the two pieces of metal do not join each other on the same plane. The welder may want two pieces of metal joined at a 90° angle to form a corner (see Figure 20-9). This weld joint is used commonly on trailers, wagons, and other square or rectangular pieces of equipment. The weld is made at the outside junction where the two pieces touch each other. Often the weld is made without a filler rod. The edges of the base metal are heated to supply the necessary molten puddle.

To make an outside corner weld on 16-gauge metal (0.134 in, or 3.4 mm), start with a number 3 tip and set both gas pres-

sures at 5 psi [0.35 bars]. Tack the base metal to form the 90° angle. Do not overlap the edges. The base metal should be positioned so that the weld joint is at a right angle to the person welding.

Hold the welding torch at right angles to the base metal and the tip at an angle of approximately 15° in the direction of travel. Now, hold the tip of the inner cone about 1/16 in [1.6 mm] above the edge of the base metal. Move the tip in a sidewise motion across the weld joint. The torch must move slowly enough so that the base metal melts to form the molten puddle. Move forward about 1/16 in with each motion. Fill the molten puddle at the end of the corner weld to ensure maximum strength of the weld.

Fillet Welds

A fillet ("fill-it") weld is sometimes called a *T joint*. It is formed by two pieces of metal welded together so that one piece is perpendicular to the other (see Figure 20-10). This weld joint is commonly used on farm equipment and food-processing machinery.

At the same time, the flame must heat both edges of the base metal and the angle where they join. To cover all this area, it may be necessary to use a tip that is slightly

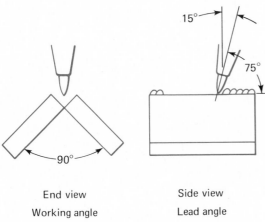

15°

75°

90°

End view

Working angle

Side view

Lead angle

Figure 20-9. Torch angles used to form the corner weld.

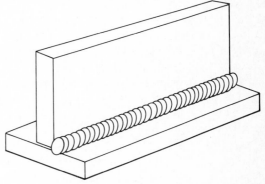

Figure 20-10. A fillet weld or T joint.

larger than what would normally be used on the metal being welded. Either the forehand or the backhand technique may be used, depending on the thickness of the metal. The welding tip is held at a 45° angle to the base metal and angled approximately 15° in the direction of travel.

Working on 16-gauge metal, select a number 3 tip and set both the oxygen and acetylene pressures at 5 psi [0.35 bars]. Check for specific manufacturer's recommendations. Tack the base metal at predetermined intervals.

The joint should be positioned at a right angle to the person welding. If you are right-handed, start on the right with the tip set at an angle of 60° to the base metal. The tip should be at a 15° angle to the direction of travel. Move the inner cone across and slightly up on the base metal to give equal heat to the entire area. Form a molten puddle in the corner of the base metal with a leg of $3/32$ to $1/8$ in [2.4 to 3.2 mm] (see Figure 20-11).

Dip the end of the filler rod into the rear portion of the molten puddle. Complete the weld joint by moving the molten puddle about $1/16$ in [1.6 mm] with each motion up and across. The weld is finished in the same manner as the butt weld. To test the weld, bend the vertical piece toward the bead. It should bend without breaking.

Lap Welds

A *lap weld* is formed when two pieces of metal overlap each other (see Figure 20-12). Since welds can now be made along two joints, the lap weld is particularly useful when strength is very important. Also, when thicker metal is used, it is possible to omit beveling if the two pieces of metal are overlapped.

Select a number 3 tip and set both gases at 5 psi [0.35 bars]. Tack the 16-gauge base metal. The weld joints are positioned at a right angle to the person welding. Hold the welding torch at a 75° angle to the base metal and the tip at a 30° angle to the direction of travel.

Move the tip across the base metal, tying the bead to the top edge of the upper piece of metal. Again, use a slight sidewise motion to produce a bead with the horizontal leg length equal to the thickness of the metal.

Move forward about $1/16$ in [1.6 mm] with a sidewise motion. This will produce a uniform ripple on the face of the bead. Completely fill the end of the weld joint as described for other welds.

The same basic backhand procedures are followed in welding on thicker metal.

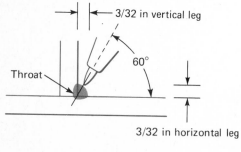

Side view
Working angle

Figure 20-11.   Torch angle used to form the fillet weld.

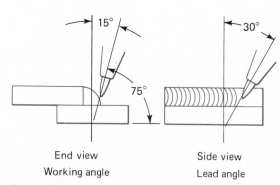

End view
Working angle

Side view
Lead angle

Figure 20-12.   The lap joint is commonly used to repair agricultural machinery.

GAS WELDING, CUTTING, AND HEATING

Fusion Welding: A Review

Today the majority of the equipment used in agricultural industry is made with metal. Much of this metal is mild steel and is joined by fusion welding. This fusion-welding process involves melting two pieces to fuse into one piece. Some types of weld joints require filler metal to add to the weld. This is called a filler rod, and it is of the same composition as the base metal. Metal preparation is a key to quality gas welding. The metal must be free from rust and other foreign materials.

The thickness of the metal determines the volume of heat required to melt the weld puddle and keep it plastic. The volume of heat is regulated by the orifice (opening) size of the tip. Tip size is designated using a number, with the lowest number having the smallest orifice. Because there are variations in the way manufacturers designate tip size, you should always check the technical chart for the tip number.

Two basic welding techniques are used in fusion welding, depending on the thickness of the metal being welded. The forehand technique is used on metal which is $3/16$ in [4.8 mm] or less. A right-handed welder will tack the joint and then begin the weld on the right side, moving across the joint. The torch is held at a 15° angle in the direction of travel. The backhand technique is less common; it is used on metal $3/16$ in thick or thicker. The operator begins on the left side and moves the torch to the right. This technique requires a 30° torch angle and produces a narrower and deeper weld.

There are four common joints which are welded. These are the butt, corner, fillet, and lap weld. Each joint has specific applications.

THINKING IT THROUGH

1. Why is metal preparation important when fusion welding mild steel?
2. Sketch the fit-up of the following joints:
   - a. $1/8''$ mild steel butt weld
   - b. $1/2''$ mild steel butt weld
   - c. 10-gauge fillet weld
   - d. $3/8''$ mild steel lap weld
   - e. $3/32''$ mild steel outside corner weld
3. What tip size should be used for each of the above joints?
4. Does the temperature of the oxyacetylene flame increase as the tip size increases?
5. Describe how working pressures are set.
6. Sketch the three types of welding flames.
7. Which of the welding flames is used for fusion welding?
8. What determines the welding technique to be used?

CHAPTER 21

BRAZE WELDING AND SOLDERING

Brazing and soldering are used to repair light-gauge sheet metal found on combines, grain drills, tractors, and many other pieces of agricultural equipment. Brazing and soldering also can be used to join two pieces of metal that are not alike, for example, steel and copper or aluminum and cast iron.

An engine mechanic working in a farm and garden center may find that the hood and grill of a lawn and garden tractor have split around the bolt holes because of improper maintenance. The choice for repairing the part is limited to (1) trying to buy new parts or (2) repairing the original parts. Because the mechanic can braze with the oxyacetylene torch, the repairs are made for less than $20. New parts would cost 4 times that amount and may take weeks for a special order. Other agricultural workers such as combine operators, tractor mechanic's helpers, and agricultural supplies/services employees experience similar problems and benefit by being able to braze and solder.

A major difference between braze welding and fusion welding is that in braze welding the base metal is not melted. The weld is a band of a brazing alloy that joins by adhesion rather than fusion. The filler metal alloy melts at more than 800°F [427°C]. The braze weld may not be as strong as a fusion weld, but it is stronger than a soldered joint and, if correctly done, has a tensile strength of approximately 55,000 psi [379.5 MPa].

Soldering is similar to brazing, but the joint is not quite as strong—25,000 to 29,000 psi [172.38 to 199.96 MPa] of tensile strength. Also, the joint must be kept at working temperatures of 300°F [149°C] or less. Otherwise, the solder may remelt, and the joint will break.

CHAPTER GOALS

In this chapter your goals are:

- To be able to identify metals that can be joined by brazing and soldering
- To prepare the base metal and braze or solder different kinds of metal to create a variety of joints
- To do these things safely

Preparing the Metal

Metal preparation is very important, because the two metals do not fuse but adhere to the weld itself. Another way to look at it is to think of the weld as a glue. When glue is used to join two things, the bond is made when the glue adheres to both pieces. By heating the base metal, the

228

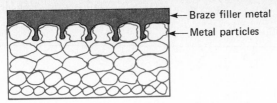

Figure 21-1.   A microscopic view of a brazed surface.

grain structure opens, allowing the braze or solder to flow in between grain structure (see Figure 21-1).

In both brazing and soldering, a flux must be used to clean the metal. Flux is a chemical that removes the oxides and other substances. It comes in liquid, paste, powder, and solid forms. Choose a flux that is most suitable for the metals being worked on.

There is more to preparing the metal than cleaning it with flux, however. The surfaces also should be scrubbed with a wire brush, a stainless-steel brush, or a medium-grade (120-grit) abrasive cloth. This removes heavy oxides, mill scale, paint, grease, and other foreign materials.

The parts to be brazed or soldered must be aligned carefully. Allow a gap of 0.003 or 0.004 in [0.08 to 0.10 mm]—the thickness of a sheet of paper—between the two pieces of metal where the braze or solder will flow and fill. A joint that is going to be brazed requires a slightly tighter fit than one that is going to be soldered because the filler metal used is more fluid.

Brazing Process

Braze welding is made at temperatures of 800°F [427°C] and above, but the temperature of the weld joint should always be below the melting temperature of the base metal. Braze welding is sometimes called *brazing* or *bronze welding*. Technically, brazing uses a copper-zinc filler rod while bronzing uses a filler rod make of a copper-zinc-tin alloy.

Advantages and Disadvantages

Braze welding requires a clean surface. Otherwise, the filler metal will not adhere to the base metal. Clean the surface with a wire brush or medium-grade (120-grit) abrasive cloth.

Many filler metals can be used for brazing. One of the most common used in agricultural industry combines 60 percent copper and 40 percent zinc. Other kinds of filler metals include nickel and chromium alloys, silver alloys, aluminum and silicon alloys, and magnesium alloys. Select a filler rod that has a melting point at least 100°F [38°C] below the melting temperature of the base metal.

The braze-welding process has several advantages. The process requires less heat than fusion welding. This means it will be faster, use less gas, and have less distortion of the welded joint. Since the process is done at temperatures below the melting point of the base metal, extensive preheating of the area around the joint is not necessary. Also, the brazing process enables dissimilar metals such as cast iron, copper, and steel to be joined.

The disadvantages of the process include the restricted operating temperatures of the joint to below 500°F [260°C]. The weld loses much of its strength at this temperature. The brazed joint should not be used on steel which is under stress, because the fatigue will cause failure of the joint. Another disadvantage is that the color of the weld is different from that of the base metals. Depending on the application, this may be either an advantage or a disadvantage.

It is important not to heat the metal too fast or to stay on one spot too long. Know the metal being brazed. Cadmium-plated metals should not be brazed, because of the dangerous fumes. Therefore, make sure that the work area is well ventilated and the space is not confining. Figure 21-2 shows the kinds of flames used in braze welding.

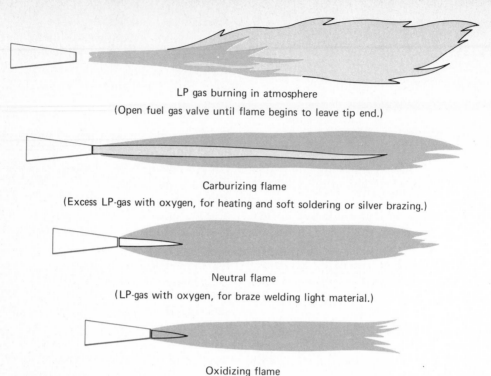

LP gas burning in atmosphere

(Open fuel gas valve until flame begins to leave tip end.)

Carburizing flame

(Excess LP-gas with oxygen, for heating and soft soldering or silver brazing.)

Neutral flame

(LP-gas with oxygen, for braze welding light material.)

Oxidizing flame

(LP-gas with excess oxygen, hottest flame about 5300° F (2927°C), for fusion welding and heavy braze welding.)

Figure 21-2. Braze welding can be done with acetylene or with a LP-fuel gas.

Brazing Sheet Metal

Sheet metal is widely used in agricultural industry. Here are just a few examples of where it is found: roofing, machinery parts, trailer and truck beds, grain elevators, truck and tractor bodies, and water tanks. It is no wonder, then, that knowing how to braze-weld sheet metal is so important to the agricultural worker.

Some advantages and disadvantages of the brazing process have already been mentioned. For example, braze welding requires less heat and is a faster process than fusion welding. But braze welding is not a good choice for metals that will be subjected to high heat (over 500°F, or 260°C). Another characteristic of the braze weld is that certain kinds of steel parts and equipment should be brazed only once. For example, mower pitman arms or other parts subject to a push-pull type of wear

may crack if they are welded more than once.

Before you begin to braze any joint, the surface of the base metal should be thoroughly cleaned. Align the parts as described previously. (See Figure 21-3 on aligning the metal for brazing a lap joint.) Heat the tip of the brass filler rod and dip it into the flux. Some flux should stick to the rod.

Use a neutral or very slightly oxidizing flame to preheat the base metal to a dull cherry red (about 1500°F, or 816°C). Don't overheat, however, or the weld may become porous. Touch the fluxed filler rod to the heated portion of the joint, and allow some flux to melt and react with the base metal. This helps to clean the base metal of oxides.

Melt a small amount of the fluxed rod until it flows freely to *tin* the joint—to

Correct　　　　　　　Incorrect

Figure 21-3.　Correct alignment is very important when brazing.

Butt weld　　　Fillet weld　　　Corner weld

Figure 21-5.　Three different kinds of joints that are brazed.

apply a light coating of the brass or bronze. When this happens, the correct temperature of the base metal has been reached. If the filler metal forms beads, the base metal is not hot enough. However, if the filler metal spreads over a large area, the surface is too hot. Maintain the proper temperature of the base metal by running the tip of the flame's inner cone over the joint at a height of about 1/16 in [1.6 mm].

As the bead is built up, continue to dip the filler rod into the flux before melting. After completing the joint, examine it. If it is copper-colored with a white smoke and has a small valley down the center of the bead, the base-metal temperature was too hot. If the edge of the bead appears to lap over the base metal, the temperature of the base metal was too cold (see Figure 21-4).

Figure 21-5 shows examples of how different kinds of joints are brazed or bronze-welded.

Braze-Welding Cast Iron

Suppose a cast-iron part (casting) on the equipment breaks and it is necessary to wait days or weeks to get a new part. Braze welding is an excellent way to fix the casting and put the equipment back to work quickly. Bronze makes a very strong weld on cast iron—often stronger than the original part.

Large castings should be preheated until they turn black, at about 400°F

[204°C]. They should be brazed while still hot. This avoids putting too much stress on the metal. The preheating can be done either in a furnace or by moving the flame over the casting before and during the brazing.

Cast iron contains large particles of carbon. A carbon film may form which could interfere with the bonding of the joint. To remove the film, carefully heat the joint to a cherry-red color before applying the flux. This is called *searing*, and it burns the carbon out of the joint surface.

The broken casting must be beveled to allow the weld to fully penetrate. The cut should be at an angle of about 30°. Also, be careful not to destroy the original size and shape of the casting. This can be avoided by leaving a shoulder of 1/16 in [1.6 mm] on the original break line at several points (see Figure 21-6).

In addition to following the procedures just mentioned, be sure to clean the metal with a wire brush or grinder. Clamp the pieces, if possible, to prevent them from moving during the welding process. As in other brazing, the joint should be tinned.

Where the castings are very thick, use

Correct　　　Too hot　　　Too cold

Figure 21-4.　A lap joint is a typical brazing job on agricultural machinery.

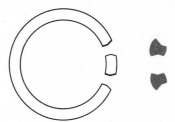

Figure 21-6.　Castings can be brazed using the oxyacetylene torch.

Figure 21-7. A three-pass weld joint.

a multipass bead, extending each bead over the previous one (see Figure 21-7).

Brazing Aluminum

There are two precautions about brazing aluminum. First, aluminum does not change color when heated. Therefore, it is very difficult to tell when the base metal has been properly heated. Second, the melting temperature of aluminum filler rods is very close to the melting point of the aluminum base metal. Most aluminum filler rods melt and flow between 1080 and 1165°F [582 and 629°C]. Most aluminum alloys flow at less than 1200°F [649°C].

As in all other brazing, it is important to thoroughly clean the base metal. Use a commercial flux. Now, using a slightly carburizing flame, heat the filler rod and dip it into solid or powdered flux. Apply some of the flux directly to the joint. Then heat the flux in the joint until the flux turns to a liquid. This happens at a temperature of approximately 950°F [510°C], well below the melting point of the metal. Continue to heat the joint carefully until the filler metal flows into the joint.

When the weld has been completed, let the joined pieces cool before moving them. The material should cool down to approximately 900°F [482°C]. Clean the weld area with water to remove all remaining flux. If flux is left on, the acid will destroy the quality of the weld.

Silver Brazing

While often called soldering, *silver brazing* is actually a brazing process. Silver brazing is used frequently in joining stainless-steel, copper, brass, or mild-steel pieces to form equipment that will be used in the food-processing industry.

When properly done, silver brazing produces a very smooth bead that resists the formation of bacteria around the weld. The silver alloy flows at temperatures of 800 to 1650°F [427 to 899°C]. Stainless steel melts above 1300°F [704°C]. This usually requires an oxy-fuel torch as a heat source.

In silver brazing, the stainless-steel base metal is first brushed with a stainless-steel brush or stainless-steel wool. Do not use a common wire brush. The two pieces should be aligned so that only 0.001 in [0.025 mm] separates them. Apply a liquid flux to the joint, brushing it on.

Begin to heat the base metal. The correct temperature has been reached (approximately 1100°F, or 593°C) when the flux becomes a clear liquid. Now apply the silver-brazing alloy, using the heat from the base metal and the flame envelope to melt it into the joint. The 45 percent silver alloy filler rod melts at 1120°F [604°C] and flows at 1140°F [616°C].

When the work is completed, allow the material to cool before moving it. Wash the joint thoroughly with water to remove any flux.

Soft-Soldering

Soft-soldering is joining two metals using very low heat. The melting temperature of the solder ranges from 361 to 640°F [183 to 338°C]. In no case does the flowing temperature of soft solder exceed 800°F [427°C]. The soft-soldering process is used to join metals that are not subjected to stress or tension. The soldered joint has a very low tensile strength. If mechanical strength is required, special joints should be used. (See Figure 21-8.)

It is important to know the metals being joined and what solder will best do the job. Solder is an alloy of tin and lead. (Sometimes small amounts of other metals may be added, but solders containing these

232

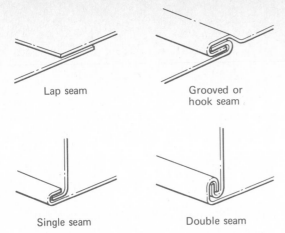

Lap seam

Grooved or hook seam

Single seam

Double seam

Figure 21-8. Special soldered joints provide added mechanical strength.

other metals are not commonly used in agricultural industry.) Solders are rated according to the percentages of tin and lead contained in them. A general-purpose solder contains equal amounts of tin and lead, and often it is referred to as a 50/50 solder. It also may be designated a number 1 solder. Other solders are 60/40, 40/60, and 30/70. The percentage of tin is always the first listed.

Solder will start to soften at a specific melting temperature and will flow freely at a slightly higher temperature (see Table 21-1).

The selection of the flux to clean the base metal is also important. Several common chemicals may be used as a flux, although different kinds of flux may be purchased. Table 21-2 lists some common fluxes.

A *hollow-core solder* is popular to use because of its convenience. This tubelike

TABLE 21-1.   Melting Point of Solders

| % TIN | % LEAD | TEMPERATURES | |
		F	C
50	50	415°	210.65°
45	55	437°	222.75°
40	60	459°	234.85°
60	40	360°	180.40°

TABLE 21-2.   Common Soldering Flux

MATERIAL	FLUX
Iron and steel	Sal ammoniac or acid
Galvanized iron	Muriatic acid
Copper and brass	Rosin, sal ammoniac, or acid

wire contains either a rosin or muriatic acid flux.

For small jobs, it is better to choose a common commercial flux. It is possible to purchase soldering kits that contain a variety of solders and fluxes. After soldering the joint, wash away the flux.

Soldering Equipment

There are three common heating tools used in soldering: (1) the *electric soldering gun*, (2) the *soldering iron*, and (3) the *torch*.

The electric soldering gun is used frequently because it is easy to operate and inexpensive. The soldering gun is available with a single or double heat range and with a wattage capacity of 150 to 230 W. The larger the number of watts, the more heat is produced at the tip. The gun should have replaceable tips for use on different kinds of metals. Before a tip is used, it must be tinned—coated with a very thin film of solder. This is done by heating the tip to operating temperature and applying solder directly to the tip. Shake the tip slightly to remove excess solder, or rub the tip on a block of sal ammoniac. A clean, damp cloth to wipe off the solder gun will aid in cleaning and tinning.

There are several kinds of torches that may be used for soldering. They use a variety of gases, including propane, MAPP gas, and butane. The oxygen for mixing with the other gas may be provided as a gas or as solid pellets. Generally the torch is more expensive than the soldering iron.

Soldering Procedure

The procedure may vary slightly depending on the soldering equipment used, but always begin by thoroughly cleaning the surface of the two metals to be joined. Use an abrasive cloth or stainless-steel wire brush, followed by flux. The two metals should be aligned so that there is a gap about the thickness of this sheet of paper, or 0.003 to 0.005 in [0.08 to 0.13 mm].

Before the solder is applied, the edges of the two metals are heated. The metals themselves then conduct the heat to the joint. Moving along the joint, apply heat, remove the heat, and apply the solder—in that order. After the solder has penetrated the joint and hardened, wash away the flux with water.

Make sure the two metals do not move during the soldering process. If the joint moves before the solder hardens, a crack could develop that would result in joint failure. If there is a possibility of the metals moving, clamp them in position.

Soldering Aluminum

It is particularly important to vigorously brush aluminum with a stainless-steel brush because aluminum sometimes looks cleaner than it really is. Aluminum may contain a heavy layer of oxide, or rust, but not show it. Aluminum does not change color when it oxidizes.

The aluminum joint is heated with a slightly carburizing flame. A large, circular motion is used, since aluminum is a good conductor of heat. The filler rod is melted by the heat of the base metal. The rod is rubbed against the hot metal to aid in breaking through the oxide film. A good commercial flux will improve the quality of the soldered joint.

After completing the joint, remove all traces of the flux with water. If all the acid flux is not removed, the metal will corrode. To further protect the joint from corrosion,

the joint may be painted or sprayed with a clear lacquer.

Soldering Copper Tubing

Many buildings now contain copper plumbing, and all the joints are soldered.

Before soldering a joint, clean the end of the tube and the inside of the fitting with an abrasive cloth. Rub until the joint is bright. Dip the tube and fitting into a flux. Align the parts. Then heat the fitting to about 650°F [343°C]. Apply the solder to the opening around the fitting. The heat will melt the solder and draw it into the joint. A small ring of solder should form around the inside of the joint. When soldering is finished, wipe the joint with a damp cloth to remove excess solder from the outside of the joint.

A silver alloy is often used in joining copper and stainless-steel tubing in food-processing equipment. Since this is a very expensive process, the worker must be very skilled at conventional soldering before trying the silver alloying process.

Safety Rules

The following safety rules should be followed in brazing and soldering:

1. Use personal safety equipment, including goggles, gloves, and proper clothing.
2. Observe approved safety practices when using compressed gases.
3. Use caution when handling an acid flux. Zinc chloride and muriatic acid give off dangerous fumes and should be used in a well-ventilated area.
4. Always braze in a well-ventilated area. The bronze may give off zinc fumes if the base metal is too hot for proper welding.

Braze Welding and Soldering: A Review

Braze welding and soldering are sometimes referred to as *hard-* and *soft-soldering*. These two processes are distinguished by the temperatures that are used to form the joint. Both use temperatures that are less than the melting point of the base metal. However, brazing (hard-soldering) is done at temperatures of above 800°F [427°C]. Soft-soldering is done with temperatures of less than 800°F and usually in the range of 361°F to 640°F [183 to 338°C], depending on which soldering alloy is used.

Metal preparation is very important. Because these techniques use adhesion rather than fusion, the area must be cleaned. Mechanical cleaning such as with a wire brush or abrasive cloth is recommended to remove paint, scale, rust, or other foreign material. A cleaning flux that removes oxides should also be used. A variety of commercial fluxes are available for special jobs.

The brazing process requires a clean surface in order for the braze or bronze to adhere to the surface of the base metal. Using a neutral flame, heat the joint to a dull-red color before the filler metal is added. If the base metal is too cold, the filler metal will form beads. If the base metal is too hot, the filler metal will "boil" and will spread over a large area. Some practice is needed to determine the correct temperature. Many common metals may be joined by using the brazing process.

Soft-soldering is used where mechanical strength is not required and the joint is not subjected to temperatures above 300°F [149°C]. The common solders are an alloy of tin and lead, with the percentage of tin listed first on the label. Three common solders are 50/50, 40/60, and 30/70. The solder is applied to the metal with a soldering gun, soldering iron, or torch as a heat source. The electric soldering gun is popular for small jobs because it is fast and economical. However, if a large amount of work is to be done or if tubing is to be soldered, a torch will work best. These torches use acetylene, MAPP gas, butane, or propane as a fuel.

THINKING IT THROUGH

1. Describe the basic differences between fusion welding, braze welding, and soft-solder.
2. What are the primary metals found in braze and bronze rods?
3. Since the brazing and soldering processes do not fuse the base metal, describe how the metals are joined.
4. Describe the preparation procedure which should be followed before brazing or soldering.
5. How can mechanical strength be increased using a soldered joint?
6. What are the primary metals used in soft-solder, and what are the common proportions available?
7. What special equipment is available for soft-soldering?
8. Describe the purpose of and the procedure for tinning.

CHAPTER 22

GAS CUTTING AND HEATING

The oxy-fuel torch also can be used for cutting, piercing, and heating all ferrous metals such as mild steel, carbon steel, and low-alloy steel.

The cutting torch is portable and can be used in the field as well as in the shop. Because the cutting torch has so many uses, nearly all agricultural workers should know how to use it correctly and safely. It is a basic tool for repairing broken and worn machinery and for remaking and modifying older equipment. The heating tip also is used to heat steel for reshaping and to remove press-fit parts.

Here is just one situation in which an agricultural worker would use an oxy-fuel cutting torch: A combine can't be used until an important bearing is replaced. But the condition of the bearing and the shaft is such that the only way to remove the bearing is to cut it and drive it off the shaft. The oxy-fuel torch can be used for this type of job.

CHAPTER GOALS

In this chapter your goals are:

- To set up the oxy-fuel torch for cutting
- To select the correct cutting and heating

tips and working pressures for specific jobs
- To heat steel for reshaping and to heat press-fit parts to aid in their removal
- To cut mild-steel plate, angle, and pipe, and to cut holes in (pierce) steel
- To demonstrate how to work safely

Setting Up Cutting Equipment

Start with the same basic equipment as for welding. Therefore, regulators, hoses, and the torch handle are set up in the same way (see Figure 22-1).

Before attaching the cutting head, inspect the connections, coupling nut, and cutting tip. Look for any signs of damage to the O-ring seals. Be sure there is no grease, oil, or dirt on these parts. If there are signs of oil or grease on the cutting head, wash it with soapy water. Dry the head thoroughly. If grease is inside the assembly, notify the instructor. The instructor will make sure the grease is removed by a service repair station.

The procedures listed below should be followed to set up most cutting torches. Check the owner's manual for the torch, however, to see if there are any other procedures one should know about.

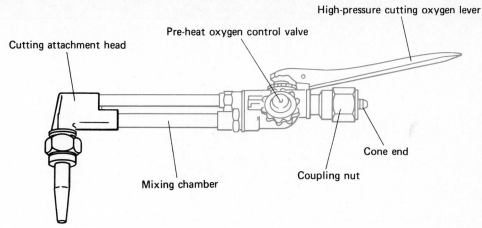

High-pressure cutting oxygen lever

Pre-heat oxygen control valve

Cutting attachment head

Cone end

Coupling nut

Mixing chamber

Figure 22-1.   A Victor cutting attachment.

Connect the cutting head to the welding torch handle, and tighten the coupling nut. Firm hand pressure is ample on most torches, since both the metal parts and the O rings create a seal. Select the right size and type of cutting tip for the job. The thickness of the metal and the type of fuel used determine the size tip. Several fuel gases, including acetylene, MAPP-gas, LP-gas, and HPG, are satisfactory for cutting. See Tables 22-1 and 22-2 as guides in tip selection.

Inspect the tip. Make sure the preheat and cutting oxygen openings are clean. If they appear dirty, use a wire tip cleaner. Be careful not to break off the tip cleaner inside the opening. Some tips are shown in Figure 22-2. Insert the tip into the torch head and tighten the retaining nut to about 20 foot-pounds (ft/lb) [27.15 N/m] of torque. This is snug but not overly tight. Check the manufacturer's literature for specific recommendations.

Set the correct working pressures on

TABLE 22-1.   Cutting Tip Series: (For use with acetylene)
Types: 1-101, 3-101 and 5-101

METAL THICKNESS	TIP SIZE	SPEED (in/min) MIN./MAX.	OXYGEN CUTTING (psig) MIN./MAX.	ACETYLENE (psig) MIN./MAX.	(kerf) WIDTH
1/8"	000	28/32	20/25	3/5	.04
1/4"	00	27/30	20/25	3/5	.05
3/8"	0	24/28	25/30	3/5	.06
1/2"	0	20/24	30/35	3/5	.06
3/4"	1	17/21	30/35	3/5	.07
1"	2	15/19	35/40	3/6	.09
1 1/2"	2	13/17	40/45	3/7	.09
2"	3	12/15	40/45	4/9	.11

Source: Victor Welding Co.

GAS CUTTING AND HEATING

TABLE 22-2. Cutting Tip Series: GPM, HPM, 1-303M
(For use with: liquid air fuel-gas, MAPP gas)

METAL THICKNESS	TIP SIZE	(in/min) MIN./MAX.	CUTTING OXYGEN-(psig) MIN./MAX.	PRE-HEAT FUEL-(psig) MIN./MAX.	(kerf) WIDTH
1/8"	000	24/28	20/25	2/5	.04
1/4"	00	21/25	20/25	2/5	.05
3/8"	0	20/24	25/30	3/5	.06
1/2"	0	18/22	25/35	3/5	.06
3/4"	1	15/20	30/35	3/6	.08
1"	2	14/18	35/40	3/6	.09
1 1/2"	2	12/16	40/45	4/8	.09
2"	3	10/14	40/45	4/8	.10

Source: Victor Welding Co.

both the oxygen and fuel regulators. These pressures vary with the thickness of the metal to be cut. Again, check the manufacturer's instructions. Check the cutting torch head, torch handle, and hose connections for leaks. This is done by using a mixture of 1 part liquid detergent and 3 or 4 parts water. When the soapy water is applied to the connections, leaks show up as bubbles. If any connection shows a leak, notify the instructor at once.

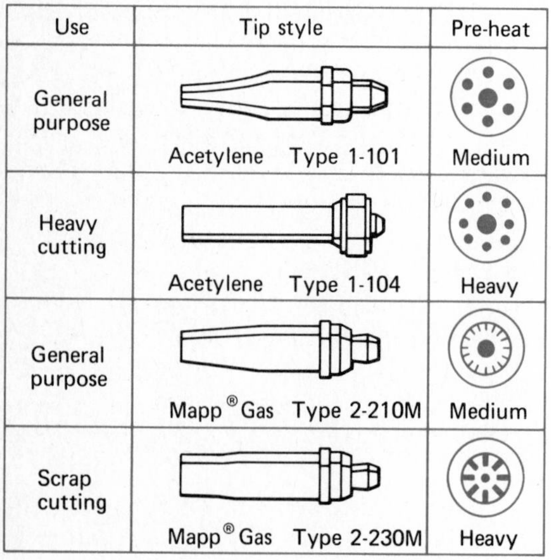

Use	Tip style	Pre-heat
General purpose	Acetylene Type 1-101	Medium
Heavy cutting	Acetylene Type 1-104	Heavy
General purpose	Mapp® Gas Type 2-210M	Medium
Scrap cutting	Mapp® Gas Type 2-230M	Heavy

Figure 22-2. Oxy-fuel cutting tips.

Open the oxygen valve on the welding torch handle all the way. Open the fuel valve on the cutting torch handle about one-half turn, and promptly ignite the gas with a spark lighter. Do *not* use a match or cigarette lighter because of the risk of injury. Increase the fuel supply until the flame breaks away from the tip. Then reduce the fuel supply slightly to allow the flame to return to the tip. There should be a proper fuel supply for the tip being used. Slowly open the oxygen valve on the side of the torch head until the preheat flame is sharp. This is the *neutral flame* used for cutting. Depress the cutting oxygen lever, and make the final adjustments with the oxygen control valve so the inner cone of the preheat flames is sharp. When the cutting lever is depressed, the pressure may drop on a single-stage regulator. Readjust the working pressure to the specifications. Figure 22-3 shows neutral flames.

Procedures for Cutting

With the cutting torch adjusted to a clean neutral flame, the torch is ready to cut.

Hold the cutting torch with both hands. If you are right-handed, the left

GAS WELDING, CUTTING, AND HEATING

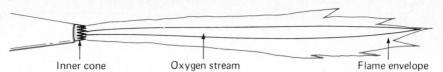

Inner cone   Oxygen stream   Flame envelope

(Acetylene with oxygen, temperature 6300°F [3482°C], proper pre-heat adjustment for all cutting.)

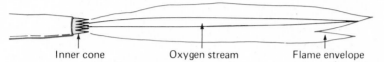

Inner cone   Oxygen stream   Flame envelope

(LP-gas with oxygen, for pre-heating 1/8 in [3.2 mm], and under prior to cutting.)

Figure 22-3.   A neutral flame is used for cutting and is adjusted by depressing the oxygen lever.

hand should support the mixing chamber of the cutting head, and the right hand should be holding the welding torch handle. The left hand should then rest on the base metal and cradle the torch so the tip is in the proper position. Direct the tip of the preheat cone where the cut will start. Position the torch so the preheat cone almost touches the surface. As the metal comes up to the kindling point or a dull cherry color, raise the cones approximately 1/16 in [1.6 mm] above the surface and depress the cutting lever. Do *not* touch the tip of the cone to the metal while cutting.

To make a square cut, hold the torch parallel to the base metal so that the head is straight up and down (at a 90° angle to the base metal). If this is not done, the cut will not be square. See Figure 22-4 as a guide. A piece of angle iron that is clamped to the metal can serve as a guide.

To make a bevel cut, the tip should be matched with the length of the cut rather than the thickness of the metal. Hold the torch head at a desired angle. Figure 22-5 shows how to make a bevel cut of 45°. If the torch is not cutting through the metal, it is probably because it is moving too fast to properly preheat the metal or the tip is too small. If the metal is fusing together behind

the cut, the torch probably is moving too slowly.

After completing the cut, turn off the fuel valve. Then turn off the oxygen valve on the side of the cutting torch head. If additional cutting is planned, it is not necessary to turn off the oxygen valve controlling the cutting orifice.

Do not lay down the cutting torch so that the cutting oxygen lever is depressed. There is still pressure on the valve. Also, don't lay down the torch while it is burning or where it might fall.

When all the cutting is finished on a job, completely turn off the cutting torch. To turn off the torch, first close the fuel valve on the welding torch handle. Then close the oxygen valve on the side of the torch head.

Close the fuel valve on the cylinder, and slowly open the fuel valve on the welding torch handle until both gauges on the fuel regulator read zero. Close the fuel regulator adjusting screw by turning it counterclockwise until it is free, but don't screw it all the way out of the body of the regulator.

Close the fuel valve on the torch handle, and close the oxygen valve on the cylinder. Depress the oxygen-cutting lever

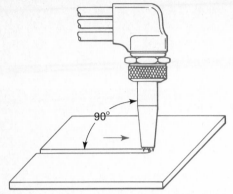

Figure 22-4. The cutting torch should be held at a 90° angle to the plate.

until both gauges on the oxygen regulator read zero. Finally, close the oxygen valve on the welding torch handle.

Roll up the hose on the welding truck or hose hanger. Do not coil the hose around the regulators or the cylinders. Place the cutting torch on the torch hanger. Do not leave the torch hanging by the hose, which would damage the hose connections.

Procedures for Piercing

Often it is necessary to make holes in, or pierce, metal. Of course, one way is to use a drill press. But this is not always possible, especially if the worker is making re-

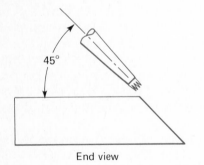

End view

Figure 22-5. Bevel cutting with the oxy-fuel torch.

pairs directly on equipment and machinery. Using a cutting torch is a good method of making holes when they don't have to be an exact size. With practice, it is possible to pierce a hole that is round and clean. However, it is difficult to make a hole with a cutting torch that will allow a perfect fit with another part.

To pierce metal with an oxy-fuel torch, set up as for cutting and hold the torch in the same manner as for cutting. The preheat inner cone should be about $1/16$ in [1.6 mm] from the surface of the base metal. Hold the torch at a slight angle to allow the molten metal to blow away.

Preheat the metal to a cherry-red color. Slowly depress the oxygen-cutting lever, and straighten the torch to a 90° angle. Move the tip in a circular motion until the desired hole size has been made. After making the hole, follow the same steps for turning off the equipment as described in the section on cutting.

Piercing generally requires more heat than cutting. Therefore, when piercing several holes, select the next larger tip than is recommended for the thickness of the metal being cut.

Procedures for Heating

It may be necessary to heat metal to reshape it or to remove press-fit parts. With some practice, this is an easy skill to master.

First select the correct-size heating tip or nozzle. Experience will help; but as a general rule, the heavier the piece to be heated, the larger the nozzle size should be. Be sure the tip does not require more fuel than $1/7$ of the cylinder capacity. Inspect the openings of the heating nozzle. If they appear to be dirty, clean them with a tip cleaner. Insert the correct-size heating nozzle into the welding torch handle.

GAS WELDING, CUTTING, AND HEATING

Tighten the retaining nut so it is snug but not overly tight.

Check the entire unit for leaks, using the solution of Ivory soap® and water. Open the fuel valve on the welding torch handle about one-half turn, and promptly ignite the fuel with a spark lighter. Increase the fuel until the flame breaks away from the tip. Then reduce the supply slightly to allow the flame to return to the tip. Slowly open the oxygen valve on the torch handle until there is a neutral flame.

For most heating jobs, the oxygen is decreased until there is a slightly carburizing flame—one that contains slightly more fuel than normal (see Figure 22-6). Keep the tips of the inner cone $1/16$ in [1.6 mm] from the base metal. Play the flame over the area to be heated. If it is a heavy piece of steel, first heat a small area to a dull-red color; then move the flame over the rest of the area, still keeping the small area a dull red.

The steel is ready to reshape when the surface to be bent is a bright, cherry red. This means it has been heated to about 1650°F [899°C]. Steel should not be bent when it is a dull red, or else it will crack.

But do not heat mild steel above 1650°F for that will reduce its strength. After the steel has been heated, close first the fuel valve to extinguish the flame and then the oxygen valve on the torch handle. Use the same procedures as outlined to completely shut off the welding torch.

Safety Rules

The following safety rules should be followed when you are cutting and heating metal:

1. Do not use oil on any part of the cutting torch or regulators. The oil will cause an explosion and fire!
2. Blow any dust out of the cylinder valves before attaching them to the regulators. This is done by quickly opening and closing the valves. The process is called *cracking*.
3. Do not use oxygen as a substitute for compressed air.
4. Do not use acetylene at pressures higher than 15 psi [1.04 bars].
5. Stand to one side when opening cylinder valves.

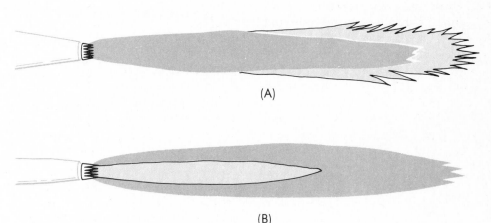

(A)

(B)

Figure 22-6. A slightly carburizing flame is used when heating. (A) shows an oxyacetylene heating flame; (B) is an oxy-LP flame.

6. Open cylinder valves slowly.
7. Make sure the oxygen and fuel passages are clear before lighting the torch.
8. Only use approved spark lighters to ignite the fuel gas.
9. Keep flames and sparks away from any materials that might catch fire or explode. Do not attempt to cut barrels or containers that were used to store flammable materials.

Gas Cutting and Heating: A Review

Many metal-working jobs may be made easier if the worker can use an oxy-fuel torch for cutting and heating. Although not as precise as the metal-cutting saw, the oxy-fuel torch can be used to cut metal to dimension. The heating torch is useful to reshape steel or to aid in the removal of rusty or frozen parts. As with the other types of gas welding, personal safety is very important.

When setting up the cutting equipment, always check the cylinders and all connections for leaks. This may be done with a commercial solution or by mixing 1 part Ivory soap® with 3 or 4 parts water.

The quality and accuracy of the cut are determined by selecting the correct tip and pressure (correctly adjusting the flame and moving the torch at the correct speed).

On occasion, it may be necessary to cut a hole in metal without the aid of a drill press. If the hole does not have to be a specific size, the cutting torch may be used.

THINKING IT THROUGH

1. Describe the correct safety procedures for setting up the oxy-fuel cutting unit.
2. Using Table 22-1, select the tip size and oxygen and acetylene working pressures for each of the following jobs:
   (a) Cutting steel plate $1/8$ in [3.2 mm] thick
   (b) Cutting steel plate $3/8$ in [9.5 mm] thick
   (c) Piercing steel plate $3/8$ in thick
   (d) Cutting steel plate $1/2$ in [12.7 mm] thick
3. Using Table 22-2, select the tip size and oxygen and MAPP gas pressures for each of the jobs outlined in question 1.
4. Describe or sketch the technique used to make square cuts and bevel cuts of 45°.

UNIT VII

TOOL REDRESSING AND CONDITIONING

COMPETENCIES

	PRODUCTION AGRICULTURE — Farmer and rancher	AGRICULTURAL SUPPLIES/SERVICES — Fertilizer applicator	AGRICULTURAL MECHANICS — Equipment setup person	AGRICULTURAL PRODUCTS, PROCESSING, AND MARKETING — Dehydrating plant worker	HORTICULTURE — Nursery worker	FORESTRY — Sawmill operator	RENEWABLE NATURAL RESOURCES — Conservation field agent
Describe correct safety procedures for redressing hand and power tools	Very Important	Very Important	Very Important	Very Important	Very Important	Very Important	Very Important
Redress common cutting tools such as knives, plane irons, and twist drills	Very Important	Very Important	Very Important	Important	Very Important	Very Important	Very Important
Redress power cutting tools such as chain saws and mowers	Very Important	Important	Very Important	Important	Very Important	Very Important	Very Important
Redress chisels, punches, and wedges	Very Important	Very Important	Very Important	Very Important	Important	Very Important	Very Important
Redress hammers, hatchets, and axes	Very Important	Very Important	Very Important	Very Important	Very Important	Very Important	Very Important
Demonstrate proper care of mechanic's hand tools	Very Important	Very Important	Very Important	Very Important	Important	Very Important	Important
Demonstrate proper care of torque wrenches	Important	Important	Very Important	Not Important	Not Important	Important	Not Important
Demonstrate the proper procedures prior to storage of hand and power tools	Very Important	Very Important	Very Important	Very Important	Very Important	Very Important	Very Important

 Very Important

 Important

Not Important

244

Beginning in the Bronze Age, humans have relied on tools to help make their work easier and better. But even primitive people recognized the need for tool sharpening, or redressing. Today workers in agricultural industry use more sophisticated tools, but tool redressing is still an important skill.

All workers need properly conditioned tools if they are to perform their work safely and efficiently. Tools that are designed to cut, such as knives, planes, or saws, should have blades that are sharp and free from nicks. Tools that are struck, such as chisels and punches, must have both the head and the point in good condition. Hammers and other striking tools require properly conditioned heads, plus handles that are safe to use. Tools should be in their best condition, not only because they will perform better but also because tools in poor condition can be dangerous to persons using them.

A survey among farmers, ranchers, agricultural service workers, nursery workers, and foresters has shown that they all need to know how to recondition cutting, striking, and struck tools.

In this unit, you will learn the importance of having properly conditioned tools, and you will be able to appreciate the statement that a worker equipped with poorly conditioned tools is not really equipped at all. Good tools are expensive. Many times equipment setup persons must furnish their own tools. It is not uncommon for a tractor mechanic to own $1500 worth of tools in a large toolbox.

At the conclusion of this unit, you also should demonstrate how to use, care for, and store such common and often-used tools as wrenches, screwdrivers, and pliers.

CHAPTER 23

REDRESSING CUTTING TOOLS

The person who has a favorite knife knows that the knife blade eventually gets dull. The longer the knife is used, the duller it gets and the more poorly it performs.

Cutting tools simply do not perform well when their blades are dull or nicked. They do not do their job properly, and they also require the person using them to work harder and less effectively than is desirable or necessary.

To *redress* a cutting tool means to keep the blade in good condition—sharp and without nicks.

CHAPTER GOALS

In this chapter your goals are:

- To recognize why it is necessary to keep cutting tools in good condition
- To compare the different methods and equipment used to sharpen cutting tools
- To apply the proper procedures used to redress common cutting tools such as a plane, twist drills, knives, shears, and mower blades

Selecting Equipment

To properly redress cutting tools, it is necessary to choose the correct equip-ment. There are generally two basic choices to make—a *power grinder* or a *whetstone* (sometimes called an *oilstone*). The choice depends on the tool being redressed and the amount of metal that must be removed in the redressing process.

Power Grinders

The power grinder is really a grinding wheel mounted on an electric motor. It can be either a small bench grinder or a larger pedestal grinder. Grinding wheels are available in different degrees of coarseness and are used at different speeds. The selection of the proper grinding wheel depends on the job to be done with the wheel.

Grit. The degree of coarseness is designated by the *grit*. The grit, in turn, is determined by the number of holes in the sieve that screened the abrasive material used to make the grinding wheel. For example, if there were a large number of openings per square inch in the sieve, then each opening would be quite small and only the finest particles of the abrasive material could pass through. This would be designated as a fine or very fine grit.

Naturally, if there were very few openings in the sieve, each opening would be rather large. The particles of abrasive

TOOL REDRESSING AND CONDITIONING

TABLE 23-1.   Grinding Stones

COARSE	MEDIUM	FINE	EXTRA FINE
10	30	70	220
12	36	80	240
14	46	90	280
16	54	100	320
20		120	400
24	60	150	500
		180	600

material passing through would be large. The result would be a coarse grit grinding wheel.

Table 23-1 shows the grit size of different grinding wheels, from a very coarse number 10 to an extra-fine number 600.

The grinding wheels with the low-number grit (very coarse) are used for rough grinding on thick steel plate. The 30-grit wheel, for example, may be used on such parts as plowshares, heavy machinery castings, and structural steel like angle iron, channel iron, and other structural pieces.

A grinding wheel with a 100 grit is used to sharpen fine cutting edges and to shape tool steel. Incidentally, it will take more time to grind when a fine wheel is used.

Speeds.   Grinding wheels are designed to operate at specific speeds. A wheel cannot be safely or effectively operated at speeds other than the ones recommended by the manufacturer. It is possible, for example, that a wheel operated above its maximum speed could shatter, sending pieces in all directions. The extreme danger of such a situation is obvious.

The speed of a grinding wheel is designated as *surface feet per minute* (sfpm). This is different from revolutions per minute (rpm) and is determined as follows:

Multiply the circumference of the wheel in inches, C, by the motor speed in revolutions per minute; then divide by 12 to obtain surface feet per minute. The equation is sfpm = $C \times$ rpm $\div$ 12. A 7-in grinding wheel has a 7-in diameter and a circumference of 22 in (diameter times 3.14). If the motor is running at 3450 rpm, the stone has a speed of 6325 sfpm ($22 \times 3450 \div 12$).

A large pedestal grinder may have a 12-in wheel operating at a speed of 1725 rpm. The speed of the wheel would be 5419 sfpm.

In using a grinding wheel, the tool rest and guards should be adjusted so that there is no more than a $1/8$-in [3.2-mm] opening to the face of the wheel (see Figure 23-1). Also, no more than 65° of the wheel should be exposed.

Whetstones.   There are several types of whetstones, sometimes referred to as *oilstones*. They include sandstones, Arkansas stones, CARBORUNDUM® stones, INDIA stones, and CRYSTOLON stones. There are several varieties of natu-

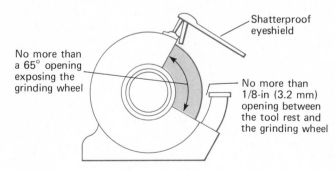

No more than a 65° opening exposing the grinding wheel

Shatterproof eyeshield

No more than 1/8-in (3.2 mm) opening between the tool rest and the grinding wheel

Figure 23-1.   Correct safety adjustments are important when using the power grinder.

ral sandstone used for redressing cutting tools. One of the most popular is the Ohio blue sandstone. The Arkansas stones are made from novaculite, a fine-grained rock found near Hot Springs, Arkansas. They are available in three grades—soft, hard, and Washita (very hard). The other stones mentioned—those with registered trade names—are manufactured in electric furnaces. They are available in grits from coarse through very fine. Some synthetic stones are also available as two-grit stones, one side being coarse and the other being fine.

These stones may be filled with oil or may require that oil or water be added as a coolant to float steel particles from the stone. Many companies have refined mineral oils designed especially for use with the stones they sell.

Procedures for Redressing Cutting Tools

All cutting tools require periodic maintenance. They can be redressed—sharpened and reconditioned—with a grinding wheel or a whetstone or both. The decision as to which to use depends on the condition of the cutting edge and which is easiest and safest for the particular situation. A good tool-grinding gauge like the one in Figure 23-2 is helpful in redressing most tools.

Redressing Plane Irons

A dull plane iron not only makes the work harder because of its resistance to cutting, but also makes uneven cuts and poor work. The cutting edge should be redressed before it becomes very dull. Usually, one can tell if the iron is dull by looking at the cutting edge. If there is a thin line, or shoulder, running along the cutting edge, it is certain that it's ready to be redressed. Also one may tell by carefully stroking the

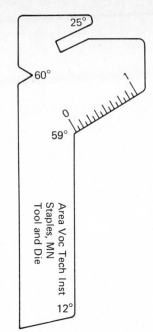

Figure 23-2. A tool-grinding gauge is necessary when redressing tools.

edge with the thumb or drawing the edge across the back of the thumbnail. A sharp plane iron will have a greater resistance as it is drawn across the nail than a dull one.

Remove the plane iron from the jack plane by removing the cap. Unscrew the cap screw, and slide the head of the screw through the slotted head in the plane iron (see Figure 23-3). If the iron has nicks or is

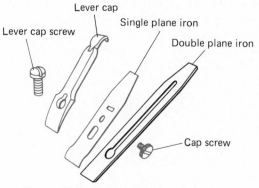

Figure 23-3. Removing the plane iron.

not square, it will be necessary to grind the edge.

It is possible to square the edge by rubbing it at right angles against a whetstone. When the plane iron is held squarely on the whetstone, the entire cutting edge should make contact. Always verify it with a try square. A grinding wheel also may be used to square an edge or remove nicks. Place the edge squarely on the tool rest. Then carefully work it across the face of the wheel until it is square and all nicks are removed. Remove only enough metal to correct the problem. To avoid overheating the iron while grinding, repeatedly dip it in light oil or water.

When the edge is square and free of nicks, it may be sharpened. Use a fine-grit grinding wheel (70 to 120 grit). Make sure the wheel is square across the face, so that the complete surface of the wheel comes in contact with the plane-iron edge. If the wheel surface is not square, ask the instructor to recondition it.

Procedure for Redressing. Adjust the tool rest so that the iron edge comes in contact with the stone at a 25° to 30° angle. If the iron is in pretty good shape, the original angle of the edge may be used as a guide.

Grip the iron so that the index finger is touching the side of the iron and firmly resting on the tool rest (see Figure 23-4). Do *not* change the position of the hand during the grinding process.

Turn the power grinder on. Slide the plane iron back and forth across the wheel. Make sure the grinding wheel is turning toward the plane iron. Keep repeating this process until there is a correct edge. Inspect the edge frequently for two reasons. First, the length of the bevel should not be greater than twice the thickness of the iron. If the bevel is longer, the plane iron may cut faster, but it also will lose the sharp edge faster. A shorter bevel won't allow

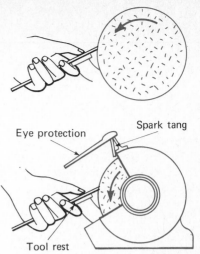

Figure 23-4. Grinding the plane iron.

the edge to be sharp enough to cut hardwoods. Second, make sure the plane iron is not overheating. Overheating can "blue" the metal and cause the plane iron to lose its edge. Cool the edge frequently with water or light oil.

Throughout the grinding, maintain the hand position as described above. Also, maintain the angle of 25° to 30°.

Make sure not to grind away any more metal than is necessary to get the desired angle on the edge. Round the corners of the edge slightly on the whetstone to prevent the plane iron from gouging or splintering wood.

Check the work frequently with a try square. The edge should be square with the side of the plane iron. If the edge is really sharp, there should be no line across the cutting edge. However, there probably will be a line, or "wire edge," on the back side of the iron after it has been ground. This can be removed by laying the back side of the iron flat against a whetstone (*not* the grinding wheel) and rubbing in a circular motion.

If a very sharp edge is desired on the plane iron, use an Arkansas-type whetstone. Lay the bevel of the plane flat against the stone. Then raise the rear of the iron by about 5°. Move the plane angle over the stone in a circular motion, applying moderate pressure. This will produce a second bevel across the cutting edge. It should be about the width of a dime.

Flip the iron over, and remove the wire edge on the back side. Apply only very light pressure and keep the plane flat against the stone.

This work with the oilstone should take about 5 minutes. If the job has been done correctly, there will be an edge sharp enough to shave the hairs on the back of the arm. Because the cutting edge may be as sharp as a razor blade, take extra care in how the plane iron is handled.

Redressing Auger Bits

The cutting action of the auger bit, which is used with a brace for boring holes in wood, is very much like that of the plane or a wood chisel. Notice how the auger bit is made. The lead screw draws the auger bit through the wood. The spurs of the auger bit notch the hole, and the cutting lips "chisel out" the wood (refer again to Figure 8-24).

To sharpen the spurs, first hold the bit upright with the tang resting on a board and the screw pointing up. Use an auger bit file or a small mill file on the inside edge of the spurs. Do not file the outside of the spur. That would affect the diameter of the auger bit and cause the hole to be smaller. Remove the same amount of metal from both spurs, but only enough to restore the knife edge to them (see Figure 23-5).

To redress the cutting lip of the auger bit, turn the bit around so that the lead screw is against the board and the tang is pointing up. File only from the upper edge of the lip. The cutting lip should have an angle of approximately 25° to the center

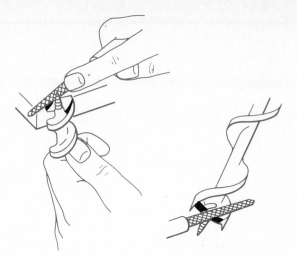

Figure 23-5.   Redressing the spurs and cutting lips of the auger bit.

line of the tang. Use the original angle as a guide.

The lead screw also may need redressing. If the threads are nicked, use a 6-in [15.2-cm] double, extra-slim taper triangular file to restore the threads. The threads must be in good condition in order to feed the bit through the lumber.

Redressing Twist Drills

Few redressing jobs have to be done more frequently than the reconditioning of the twist drill. Unfortunately, the redressing attempts often prove unsatisfactory. The main reason the redressing may not come out right is that the worker forgets the relationships of three angles.

The cutting lips should have a 59° angle with the longitudinal axis of the twist drill. The lips must be exactly equal in length (see Figure 23-6).

Also, the contour of the surface behind the cutting edge—called *heel clearance*—should be at an angle of 12° to the lip (see Figure 23-7). Too much heel clearance will cause the twist drill to cut

TOOL REDRESSING AND CONDITIONING

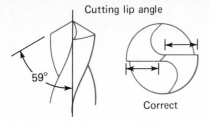
Cutting lip angle

59°

Correct   Incorrect

Figure 23-6. The cutting lips of the twist drill must be equal in length.

too fast. This will result in burning or chipping of the cutting edge. Too little heel clearance will cause the heel of the twist drill to ride against the metal. This results in excessive pressure on the twist drill and overheating.

The chisel point of the twist drill should intersect with the cutting lip to form a 135° angle (see Figure 23-8). It is the chisel point that digs into the metal and allows the cutting lip to make contact with the metal.

Procedures for Redressing Twist Drills. Keeping the three angles in mind and following the directions and the illustrations, you should be able to redress the twist bit without too much difficulty.

Begin the redressing process by holding the twist drill flute with the left hand. Hold the shank with the right hand. This is important because the twist drill is designed to turn and cut clockwise. Before starting the grinding wheel, hold the twist drill parallel to the floor or tool rest, with

the shank on the left, forming a 59° angle to the face of the grinding wheel. Advance the cutting lip toward the wheel, keeping the cutting lip parallel to the floor and the wheel. Now, lay the left index finger on the tool rest. If the angles appear to be correct, press the flute firmly against the index finger with the thumb.

Turn the grinder on with the right hand. Holding the shank with the right hand, lower it slightly and make contact with the face of the wheel. This should produce a short, sweeping motion that will result in the correct heel clearance. Continue to grind the cutting edge with short strokes. Cool the bit in either oil or water after five or six strokes, but do not release the grip on the flute with the left hand.

When one edge has been redressed, rotate the bit so that the opposite flute "fits" into the impression left on the left index finger by the flute. This "fit" should produce the same angle for the other cutting edge. Using the same technique, grind the other cutting lip.

After redressing both lips, inspect them to be sure they are exactly the same

Heel clearance angle

12°

Figure 23-7. The heel clearance angle regulates the feed rate of the twist drill.

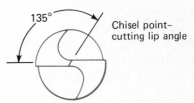
135°   Chisel point–
         cutting lip angle

Figure 23-8. The chisel point should form a 135° angle with the cutting lip.

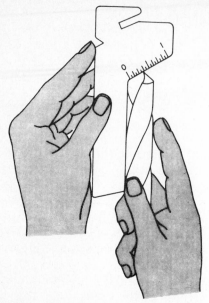

Figure 23-9. The three critical angles of the twist drill can be checked with a tool grinding gauge.

length and that the chisel point makes a 135° angle. Check all three angles, using the tool-grinding gauge (see Figure 23-9). To measure the length of the cutting lips, use a tool-grinding gauge similar to the one in Figure 23-2.

Redressing Knives

Anyone who uses a dull knife is almost surely going to be cut accidentally. The reason is that extra pressure needs to be exerted when the knife is dull, and the extra pressure may cause the knife to raise up or slip. When that happens, the user is bound to be hurt.

It takes very little effort to keep a knife properly sharpened. Usually it is enough to use a whetstone. However, if the knife blade is very dull or nicked, it may be necessary to use the grinding wheel. If so, select a stone that has a fine grit and is designed for fairly slow speeds.

Procedures for Grinding the Blade. Before turning on the grinding wheel, place the blade flat against the face of the stone. Then raise the back of the blade about 5°. This should produce an angle on the blade of about ¼ in [6.4 mm], or twice the thickness of a quarter. Holding the knife in the left hand, let the right index finger rest on the tool rest with the wheel turning toward the blade. Practice moving the blade slowly back and forth across the wheel, using light to moderate pressure against the stone. After getting the feel of it, turn on the grinder and repeat the same procedure. Take off only enough metal to remove nicks. Be sure to cool the blade frequently with water or oil.

When grinding is completed on one side of the blade, change hands and repeat the redressing on the opposite side of the blade.

Procedures for Whetting the Blade. When the grinding is completed, use a whetstone for the fine sharpening of the blade.

To *whet* the knife, place the blade flat on the face of the stone. Then elevate the back of the blade slightly. Using a sweeping motion, apply moderate to firm pressure, with the cutting edge of the blade leading as if the knife were cutting into the stone. Make sure the entire blade surface comes in contact with the stone. When one stroke is completed, turn the blade over and follow the same process with the other side of the blade (see Figure 23-10). Repeat the whetting process until there is a fine cutting edge. To test the blade's sharpness, carefully feel the edge with the thumb. A safer test is to draw the knife across the edge of a sheet of paper. The knife should cut the paper without tearing it.

Procedures for Stropping the Knife. To put the finishing touches on the knife, it

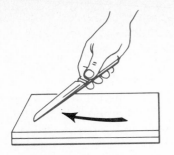

Figure 23-10. Procedures for whetting the knife blade.

may be desirable to "strop" or "steel" the blade. The two processes are similar since they do not remove metal by grinding. Instead, the microscopic "sawteeth" are aligned.

Use a piece of leather to strop the blade. Hold the blade flat against the surface, and sweep across the leather strop with the cutting edge of the knife trailing. Turn the blade over, and strop the blade in the opposite direction. The process aligns the metal particles on the blade edge and gives an extremely sharp edge.

A *steel* is an instrument that is held upright in the left hand. Holding the edge of the blade at a 20° to 45° angle to the surface, bring the blade across the steel with the cutting edge of the blade leading. Repeat the process for the other side of the blade. Usually only a few strokes on both sides are necessary to produce a fine edge (see Figure 23-11).

Redressing Axes and Hatchets

If the cutting edge of the ax or hatchet is nicked, begin by grinding the edge with a 90- to 120-grit grinding wheel. The original angle on a cutting ax is 35° to 40°, and the angle on a splitting ax is approximately 60°. The facet of the cutting edge should be 5/8 to 3/4 in [15.9 to 19.1 mm] wide, tapering off into the face of the ax.

Using proper eye protection, turn on the grinder and move the facet of the ax or hatchet across the stone with a slightly rocking motion. This is done to prevent the facet from being hollow-ground or concave. Rather, the facet should be slightly convex. Be sure the grind wheel is turning toward the ax. Remove only enough metal to get rid of nicks. Turn the head over and repeat the process.

When the grinding is complete, use a coarse whetstone or mill file to obtain a fine edge. Don't forget to remove the wire edge.

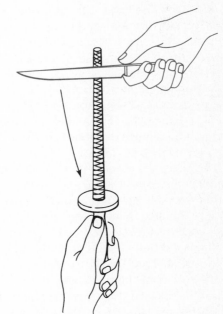

Figure 23-11. Using the steel to sharpen the knife.

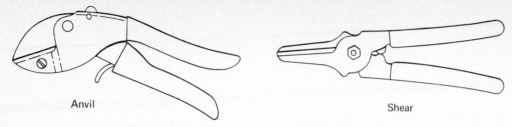

Anvil

Shear

Figure 23-12.   Two basic types of pruning shears.

Redressing Pruning Shears

There are two basic types of pruning shears: the *anvil* and the *shear*. The anvil type consists of one cutting blade working against a soft metal plate (see Figure 23-12).

Since the anvil-type shear has only one cutting edge, it can be sharpened much like a knife. If necessary, use a fine-grit wheel or a flat mill file to restore the edge and remove nicks. Do not remove more metal than is necessary. Otherwise, the life of the shear will be shortened. Hold one side of the blade at a slight angle against the grinding wheel so the wheel is turning against the blade. Apply moderate pressure and move the blade across the face of the wheel, starting with the point. Inspect the blade after every stroke. Repeat the process with the other side of the blade.

After the grinding is done, whet the blade. For small shears and those with curved blades, use either a round whetstone or a gouge slip. Both allow a curved surface to be whetted without nicking the blade (see Figure 23-13).

The shear-type blade is more difficult to redress because the blades ''scissor'' together and depend on a close tolerance between the two blades for the cutting. Always try to maintain the original angle of the blades (about 70°). If the blades are nicked, grind them on the wheel, using a fine-grit stone. Move the blade across the wheel, starting with the point. Inspect the edges after every stroke. Be sure not to close the shear during the grinding pro-

cess, because a wire edge forms on the inside of the blades.

When all nicks are removed from the blades, it is time to whet them. Holding the blades open, first place the angle against the whetstone. Then, with a sweeping motion, move the blade across the stone, with the edge leading. Make sure the entire edge of the blade comes in contact with the stone. Continue this process until a fine edge is produced. Then, carefully lay the inside face of the blade on the stone. Using light pressure and a circular motion, remove the wire edge.

Repeat the same procedures for the other blade. When both blades have been

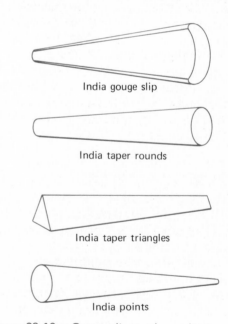

India gouge slip

India taper rounds

India taper triangles

India points

Figure 23-13.   Gouge slips and round stones used for sharpening curved blades.

TOOL REDRESSING AND CONDITIONING

redressed, check the tolerance between the blades. Some shears may be adjusted so the blades maintain the proper tolerance.

Redressing Mower Sickles

It is rather time-consuming to sharpen the various knife sections of a mower, swather, or combine sickle. And some persons think that, all things considered, the worker is better off simply replacing dull and nicked sections.

However, the sections may be sharpened. To sharpen them, it is best to use a special double-beveled grinding wheel. This wheel allows one side of the first section to be ground while grinding the opposite side of the next section.

Be sure to adjust the tool rest properly, as previously directed in this chapter. And do not forget to use proper eye protection. The mower sickle is long, and help may be needed to hold it for the grinding.

Hold the back of the sickle and position the knife sections so that they make contact with the grinding wheel near the point. Move the sickle back and forth across the face of the wheel to prevent overheating the sections. Do not grind sections more than necessary to remove nicks and restore the original angles. After doing two sections at a time, move on to another two sections (if all sections need work).

Redressing Rotary Mower Blades

Rotary mower blades operate at very high speeds—19,000 ft/min [5795 m/min] or 215 mph [346 km/h]. Because of the effects of centrifugal force, it is very important that the blade remain balanced.

Remove the blade from the mower, noting which side is on top. The bevel should be on top. Using a medium-grit grinding stone, position the blade on the tool rest so that one end of the blade contacts the wheel and the grinding action is toward the blade. Check the original angle (30° to 40°), and try to maintain that angle during the grinding.

To grind, move the blade across the wheel. It is important to count each stroke to make sure both ends of the blade will be in proper balance later. Inspect the blade after each stroke. When the nicks are removed, reverse the blade and grind the opposite end with the same number of strokes. After the grinding is done, it is advisable to check the blade on a balance point (see Figure 23-14).

Safety Rules

The following safety rules should be followed in redressing cutting tools:

1. Don't use a grinding wheel that has been worn to less than half of its original diameter. When the wheel is worn beyond one-half its diameter, there is a greater danger of its shattering.
2. Always wear protective glasses. In fact, most state laws require persons using grinding wheels to wear not only safety glasses but also a full-face shield as protection against hot sparks.
3. Be careful where the sparks are directed. They should not be directed toward other persons or into materials that may be combustible.
4. Wear long-sleeved garments and leather gloves. This is especially im-

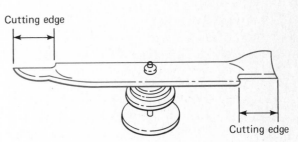

Cutting edge

Cutting edge

Figure 23-14. Check the rotary mower blade on a balance point.

portant when you are doing heavy grinding.

5. Never try to grind with the side of a wheel. Such force may cause the wheel to break.

Redressing Cutting Tools: A Review

Properly conditioned tools are necessary to do a good job. When improperly conditioned, even the most expensive tools don't work satisfactorily. Almost all redressing work is performed with an electric grinder or whetstone. Both the grinding wheel and the whetstone are available in a variety of degrees of coarseness, called grit. The grit is specified using a number from 10 to 600. The lower the number, the coarser the grinding wheel is. Grinding wheels are also designed for specific speeds. This speed is designated as surface feet per minute. A fine edge may be produced using a whetstone as the last step in redressing cutting tools.

A plane iron should be redressed with a power grinder only if the iron is extremely dull or chipped. The intersection of the cutting edge and the plane iron should form a 25° to 30° angle. A tool gauge should be used to check the angle. The plane iron may have a second angle applied to the cutting edge by using a whetstone.

Auger bits may require periodic redressing. The inside edge of the spur should be sharpened with an auger bit file or a small mill file. The cutting lip should be restored to the original 25° angle by filing on the upper side of the lip. Finally, the thread of the lead screw should be examined and restored if necessary. A special triangular file should be used.

Twist drills are a common redressing job in the agricultural shop. They should be reconditioned before they become extremely dull or chipped. The process requires only a few minutes and will save both time and energy. Three angles must be maintained. The chisel point aids in starting the hole and should form a 135° angle to the cutting lip. For general-purpose drilling, the cutting lip should form a 59° angle to the center line of the twist drill. Also, the heel of the cutting lip should have a 12° clearance. This allows the cutting lip to cut at the proper speed and pressure.

Knife edges may be redressed using a whetstone. Position the blade so a cutting angle of approximately 25° is formed on the blade. Move the blade across the whetstone with the cutting edge of the blade leading. Reverse the stroke on the opposite side.

THINKING IT THROUGH

1. What is meant by the term *grit* and why is it important in redressing cutting tools?
2. What is the common range of grit used in agricultural industry?
3. How is the speed of the grinding wheel designated?
4. How is the speed of the grinding wheel determined?
5. What is the angle of the cutting edge of the plane iron?
6. What three angles are found on the twist drill?
7. List the procedures used to redress a knife.

TOOL REDRESSING AND CONDITIONING

CHAPTER 24

REDRESSING TOOLS THAT STRIKE OR ARE STRUCK

As important as it is to keep cutting tools sharp and in good condition, it is equally important to keep tools that strike, such as the machinist hammer, and are struck, such as the chisel, in top working condition. The worker who does not keep tools sharp or otherwise in good repair is risking more than doing a poor job. That worker also risks injury.

In redressing (reconditioning) tools that strike or are struck, the same kinds of grinding wheels and whetstones are used as described in the preceding chapter.

CHAPTER GOALS

In this chapter your goals are:

- To redress striking tools such as hammers
- To redress tools that are struck, such as chisels and punches

Redressing Tools that Strike

The striking tool that most agricultural workers use, regardless of their career field in the industry, is the hammer. The worker needs to know how to recondition both nail hammers and ball peen hammers. *Reconditioning* a hammer means repairing or replacing the handle as well as sometimes reconditioning the hammerhead.

Procedure for Redressing Nail Hammers

Nail hammers are designed specifically for driving or pulling common unhardened nails and ripping apart or tearing down wooden structures. The striking face is heat-treated to produce a hardened surface, and the steel directly behind the face gradually decreases in hardness. Because of this design, the striking face of the nail hammer should not be reground, welded, or heat-treated. If the face shows signs of cracking, chipping, or mushrooming, the hammer should be discarded.

Handles should be kept tight in the head and be free of splinters or cracks. Handles that show damage should be replaced. To replace a wooden handle, clamp the head in a vise and use a hacksaw to cut the original handle from the head just behind the hammer eye. The hacksaw is used because of the possibility of striking the hammerhead's metal wedge.

Using a 1/2-in [12.7-mm] twist drill, a vise, and drill press, drill through the handle remaining in the hammer eye. Again, a twist drill is used because of the possibility of striking a metal wedge. Drive the remaining handle from the hammer eye, using a punch or a chisel. Inspect the eye for possible damage.

Select a replacement handle that is equal in size and quality to the original. A hickory-wood handle is used most often. The eye end of the handle should be dressed by using a wood rasp or a surform to fit the hammer eye. The handle usually has a slot cut into the eye of the handle for the wedge. Do not reduce the size of the handle any more than necessary. Fit the hammerhead to the handle by hand-pressing the handle into the eye. Complete the fit by sharply striking the end of the handle with a wooden mallet or striking the handle on a wooden surface. Do not drive the handle directly into the hammerhead because you may split the handle.

When the handle is properly fitted to the head, cut off the excess handle above the hammer eye. Using a metal or wooden wedge, drive the wedge into the slot to secure the hammerhead. A second wedge should be used at right angles to the first. Using the wood rasp or a sander, carefully form or burnish the end of the handle to produce a professional appearance. The hammer is now as good as new.

Procedure for Redressing Ball Peen Hammers

There are a variety of machinist's hammers that are commonly referred to as *ball peen hammers*. Most are bell-faced; that is, they have a slightly convex or rounded face (see Figure 24-1). The remainder have a flat face. All are characterized by a round knob at the opposite end of the face—the *peen*.

To redress the face of a ball peen hammer, use a flat bastard file to remove nicks and dents. If the condition of the face necessitates grinding, select a medium-grit stone. The grinding should be done at slow

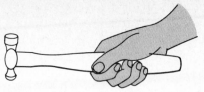

Figure 24-1.   A ball peen or machinist's hammer.

speeds—2000 to 3000 sfpm. Be careful not to overheat the face. The bell-faced hammer should have a 45° angle radius of $3/16$ to $1/4$ in [4.8 to 6.4 mm] around the edge of the face.

Ordinarily, the peen does not need redressing. However, a mill file can be used to smooth the surface of the peen.

One of the more common redressing jobs is to replace the hammer handle. It is important to select the right-sized handle. In replacing the handle, follow the procedures as outlined in the section dealing with nail hammers.

Redressing Tools that Are Struck

The principal tools that are struck which are used in agricultural industry are the various chisels, punches, and wedges. Each one requires a different kind of redressing to keep it in best working shape.

Procedure for Redressing Wood Chisels

A *wood chisel* is made from a single piece of tool steel. It is fitted with a wooden or plastic handle. It is designed to remove wood when struck with a mallet or when pushed by hand. The wood chisel works much like a crude plane. The edge of the chisel is shaped like the edge of the plane iron. The usual blade width of a wood chisel ranges from $1/2$ to $1 1/4$ in [12.7 to 31.8 mm].

If the chisel is properly cared for and used correctly, it may be redressed several times with a whetstone before it needs to be ground on a grinding wheel. If the cutting edge is badly nicked, it will be necessary to go directly to the grinding wheel.

The angle of the edge should be 25° to 30° for most work. When the chisel is used for hardwoods, such as oak and maple, it may be desirable to have a slightly longer bevel or a smaller angle. This would make the edge sharper. For some very soft woods, such as fir or pine, it may be desirable to use a chisel with an angle of 30° to 35°. By all means, *do not* use the wood chisel for cutting metal. The chisel is not properly tempered (hardened) for cutting metal.

To begin, make sure the edge of the chisel is square, that is, that the edge makes a 90° angle with the shank. A try square can be used to determine if the edge is square. If it is not, use the grinding wheel until enough metal has been removed to restore the squareness.

Now the edge can be redressed. With the grinder off, check the original angle of the chisel edge. To do this, place the edge flush against the tool rest on the grinder. Use the left index finger to guide the chisel edge as the shank is lowered or raised to find the correct angle.

Turn the grinder on and carefully work the edge of the chisel across the wheel, maintaining the original, or desired, angle. Dip the chisel in water or light oil frequently to prevent overheating and to avoid losing the original temper of the steel. Do not grind any more metal than is necessary to remove nicks and restore the original angle. Again check the edge with a try square to make sure the cutting edge is still flush with the shank.

To sharpen the edge, use an Arkansas-type whetstone or a manufactured stone filled with oil. The angle of the edge should be flat against the stone. Use a circular motion with moderate pressure. When sharpened, the bevel should be straight or slightly concave.

The head of an all-steel chisel (without a handle) may also require redressing. The head should be square with the shank (see Figure 24-2), and the crown radius should

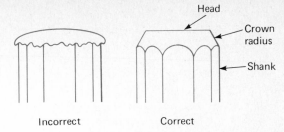

Figure 24-2. Redressing the head of a wood chisel.

taper away from the head as shown in Figure 24-1. Square the head or grind the crown radius to restore it to original condition. However, if the head is badly chipped or "mushroomed," it is better to discard the chisel.

Procedure for Redressing Cold Chisels

A variety of cold chisels are available for specific jobs. Each has its unique shape, and all are designed to cut cold metal. Also, all are forged from tool steel. Their heads and cutting edges have been heat-treated. The different kinds of cold chisels most often found in the agricultural shop are the flat, cape, diamond point, and the round nose.

The correct angle for a flat cold chisel may vary from 55° for soft metals such as brass to 90° for very hard steel. A chisel for general use should have a 70° angle (see Figure 24-3).

If the flat cold chisel has no nicks, redress the facet by using a flat bastard file. Only a few strokes are required to recondition the cutting edge. If the cold chisel is nicked, probably it should be ground on a medium-grit stone. Moderate pressure

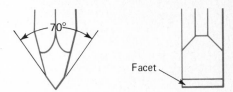

Figure 24-3. The angle used to redress a flat cold chisel.

Cape cold chisel   Round nose   Diamond point

Figure 24-4. Angles that are commonly used to redress other types of cold chisels.

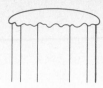

Figure 24-6. A cold chisel with a mushroomed head.

should be applied. The cutting edge should be slightly convex, as shown in Figure 24-2. Do not overheat the edge. Dip the facet in either water or light oil to cool.

To sharpen the flat cold chisel, hold the chisel at a 70° angle and move the facet across the face of the grinding wheel with the rotation toward the chisel.

The other cold chisels are redressed in a similar way. Of course, the angles may be different in each. Always sharpen to the original angle (see Figure 24-4).

Procedure for Redressing Hot Chisels

As the name *hot chisel* implies, these chisels are designed to cut and shape red-hot steel or rods. The hot chisel generally weighs 2 to 5 lb [0.9 to 2.3 kg] and is fitted with a 16-in [40.6-cm] hickory handle. The cutting edge, or bit, is between 2 and 2 1/4 in [5.1 and 5.7 cm] wide. When used, the hot chisel is struck with a heavy blacksmith's hammer.

Ordinarily, one can use a mill file or flat bastard file to redress the chisel. However, if the bit needs to be sharpened, use a medium-grit (90) grinding stone. The bit should be sharpened to a 60° angle (see Figure 24-5). When you are grinding the

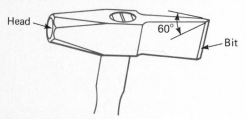

Head   60°   Bit

Figure 24-5. Redressing the hot metal cutting chisel.

bit, position the chisel so that the grinding wheel turns toward the body of the chisel. This will direct heat away from the bit. Again, be sure to cool the bit frequently to avoid overheating.

If the chisel head is "mushroomed" or in poor shape (see Figure 24-6), discard the chisel.

Procedure for Redressing Punches

There are different punches for specific jobs. The most common hand punches are the center, prick, aligning, starting, pin, and drift punches. If the punches are properly used, they seldom need redressing.

Punches are forged from tool steel and generally are from 1/4 to 1 in [6.4 to 25.4 mm] in diameter. The head and the point are heat-treated. The point of the chisel is hardened to hold the cutting edge and the head is annealed to resist the impact of striking. If the head or point of a punch has been badly damaged, it should be thrown away.

The *center punch* is used to mark the location of a hole to be drilled. Therefore, it is important that the point be concentric (have a common center) with the shank of the punch. The point is ground to a short taper point with an angle of 60° (see Figure 24-7). It requires care and experience to correctly grind a center punch by hand.

The *aligning punch* is a long, tapered instrument used to align heavy castings and other bolt holes. It should be sharpened to a 60° angle very similar to the center punch. The aligning punch is not driven by a hammer and is not used as a center punch.

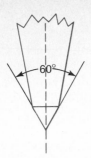

Figure 24-7. A 60° angle is used to redress a center punch.

The *prick punch* is small and is used to make small marks or reference points on materials to be worked on. It is ground to a sharp point, usually at a 30° angle (see Figure 24-8). Grind the prick punch using short and moderately light strokes. This will prevent overheating of the point.

The *starting*, *pin*, and *drift* punches all have a 90° (flat) end and are used to drive pins, rivets, and bearings (see Figure 24-9). The punch is struck with quick, sharp blows. To redress the flat point, grind the point so that it is square with the shaft and all nicks are removed. Grind the head and crown radius to restore them to original condition. As with the other tools, if the head or point is badly damaged, it should be discarded.

Procedure for Redressing Wedges

Along with the ax, workers use a *wedge* to help split logs or lumber. Wedges are struck generally with a heavy sledge hammer (maul) or the face of the single-bit ax. The wedge acts on the mechanical princi-

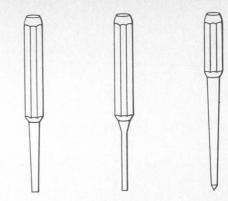

Starting punch   Pin punch   Aligning punch

Figure 24-9. Punches are used to align metal parts and to drive and remove pins.

ple of the inclined plane and is shaped like a ramp (see Figure 24-10).

Wedges come in a variety of sizes and types of metals. Some are die-cast magnesium alloy and are difficult to redress. Others are drop-forged cast or tool steel. These may be reconditioned with a flat bastard file. The file is used to sharpen the cutting edge. The edge should be sharpened to a 60° angle.

The head of the wedge should be maintained at a 90° angle, with a 3/8-in [9.5-mm] crown radius to concentrate the

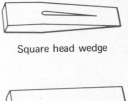

Square head wedge

Stave wedge

Oregon splitting wedge

Figure 24-10. Many wedges can be redressed with a flat bastard file.

Figure 24-8. The prick punch is redressed to form a 50° point.

force toward the center line of the body of the wedge.

Redressing Tools That Strike and Are Struck: A Review

Striking tools, such as the nail and machinist's hammer, and struck tools, such as the chisel and punch, must be in good condition to perform the job for which they were designed. Also, tools that are not redressed properly are a hazard to the worker.

The nail hammer is designed for driving or pulling unhardened nails and for ripping apart lumber. Because of the heat treatment of the head, the nail hammer should never be struck against a hard object or redressed.

Handles should be kept in proper condition and may be replaced when necessary. Because of the driving force, the handle must be kept tight by using either wooden or metal wedges.

The ball peen hammer may be redressed by using a flat bastard file to remove any nicks or dents. If a grinding wheel is used, care must be taken not to overheat the face. Overheating changes the hardness and may cause the head to fracture.

Struck tools require redressing of both the head and the point. The head should be square with the shank and have a crown radius. This radius directs the force of the blow through the center of the struck tool and lessens the chance of injury to the worker. If the head is chipped or mushroomed, the tool should be discarded. The wood chisel is redressed in much the same way as the plane iron. The angle of the blade should be approximately 25° to the center of the shank. If the edge is chipped or extremely dull, a fine-grit grinding wheel should be used to restore the original angle. The whetstone produces a very sharp edge.

Cold chisels are designed to cut cold metal. For general use, the flat cold chisel should have a 70° angle, and the facet should be slightly convex. Care must be taken to prevent the cutting edge from overheating.

The center punch should have a point which is tapered at a 60° angle and concentric to the shank. The starting, pin, and drift punches all have a 90° end.

Wedges may be redressed by using a flat bastard file. The cutting edge should be sharpened to a 60° angle, and the head should be maintained at a 90° angle with a $^3/_8$-in [9.5-mm] crown radius.

THINKING IT THROUGH

1. Outline the steps necessary to replace a hammer handle.
2. How should the head of a ball peen hammer be redressed?
3. What should be done with a mushroomed chisel?
4. What angle should be used for a general-purpose cold chisel?
5. What should be done to redress a punch?
6. How should a wedge be redressed?

TOOL REDRESSING AND CONDITIONING

CHAPTER 25

TOOL CARE

In agricultural industry today, workers depend on machinery and equipment to get many jobs done, and they rely on good tools to help them maintain and repair the machinery and equipment.

The worker on a wheat farm, for example, uses tools to keep the diesel tractor and tillage equipment in good condition. A greenskeeper at a golf course uses tools to maintain the small gasoline tractor used to mow the roughs and fairways on the course.

Because of agricultural workers' heavy reliance on tools, it is important that they know how to use and care for them. Good tools are expensive. When workers do not take care of the tools, they are wasting money as well as risking a poor quality job.

CHAPTER GOALS

In this chapter your goals are:

* To identify by name and use the common hand tools found in the agricultural shop
* To properly care for and redress screwdrivers
* To properly torque—twist tight—a nut or stud bolt to specifications
* To identify the correct sizes and uses of sockets and socket wrenches

Using and Caring For Screwdrivers

The screwdriver is a common tool that most persons have used—and abused—often. The screwdriver is available in five types of blades, but none of them is meant to be used as a chisel or pry bar.

Standard Screwdrivers

The *standard screwdriver* has a straight bit that fits a screw with a slotted head. It is classified by size according to the combined length of the bit and the shank, or blade (see Figure 25-1). The standard screwdriver is available in a stubby size—1 in [2.5 cm]—and the following sizes: $2\frac{1}{2}$, 3, 4, 5, 6, 8, 10, and 12 in [6.4, 7.6, 10.2, 12.7, 15.2, 20.3, 25.4, and 30.5 cm]. The width and thickness of the bit are usually in proportion to the length, but it is possible to purchase long screwdrivers with a narrow and thin bit. In any case, the width of the screwdriver bit should match the width of the screw slot as nearly as possible. Otherwise there may be trouble turning the screw, or the bit may damage the screw slot, making it impossible to turn.

If the screwdriver is improperly used to pry things open or off, the blade may bend. If this happens, the bit will no longer be concentric with the handle. This will cause the bit to slip out of the screw slot as the screwdriver is turned. The screw slot

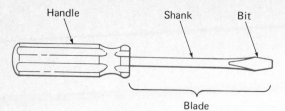

Figure 25-1. The standard screwdriver.

and the area around the screw head may be damaged.

When the wrong screwdriver is used or the screwdriver is in poor condition, the user may be injured. A screwdriver that slips and gouges the screw head also can slip and gouge the worker.

If the bit becomes chipped or slightly twisted, it may be redressed on a grinding wheel, with a medium-grit stone. Grind the tip until it is square with the shank and nicks are eliminated. Be careful not to overheat the bit, because it is made of high-carbon tool steel and may lose its temper (hardness). Dip the bit into either oil or water to cool it.

Try to grind the bit to its original thickness. Never grind the bit to a point, and do *not* try to make the edge sharp. Check the thickness and width of the bit after grinding it by inserting it into a screw slot of appropriate size. The bit should fit the slot snugly (see Figure 25-2). If the bit has been tapered, it will tend to rise out of the slot and damage the screw head and possibly the worker.

There is a limit to the number of times a standard screwdriver can be redressed. When it is no longer possible to restore the bit to its correct width and thickness, the screwdriver should be discarded.

Phillips Screwdrivers

The *Phillips screwdriver* has a cross-shaped bit with a tapered end (see Figure 25-3). It also is classified according to the length of the blade and the number size of the head. It is commonly available as a stubby—1 in [2.5 cm]—and in sizes 4, 6, and 8 in [10.2, 15.2, and 20.3 cm]. As with the standard screwdriver, it is important that the Phillips bit fit the screw slot exactly. The Phillips screwdriver ordinarily requires more downward pressure by the worker than the standard screwdriver. Therefore, it is especially important that the screwdriver fit. Otherwise it will slip and cause damage or injury. It is a good idea to brace yourself when using the Phillips screwdriver.

To redress the Phillips screwdriver bit, use a small, three-cornered file to square the shoulders of the bit. Take a

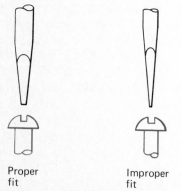

Proper fit   Improper fit

Figure 25-2. When redressing the standard screwdriver, the blade must fit the slot of the screw.

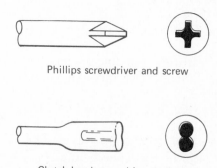

Phillips screwdriver and screw

Clutch-head screwdriver and screw

Figure 25-3. Two special types of screwdrivers.

TOOL REDRESSING AND CONDITIONING

screw with the Phillips-shaped slot and use it to test the fit of the bit. Do *not* sharpen the end of the bit to a point.

Clutch-Head Screwdrivers

The *clutch-head screwdriver* (see Figure 25-3) has a bit shaped similar to the figure 8. The clutch-head screwdriver generally is used for metal fasteners on sheet metal. The shape of the bit and the way it fits the head of the fastener help to prevent the screwdriver from slipping out of the slot.

If redressing is necessary, use a small, round file to put the bit back into its original shape. The end of the bit can be squared with a flat file.

Offset Screwdrivers

The *offset screwdriver* (see Figure 25-4) is a handy tool in the shop. It is used when it is impossible to employ a standard screwdriver because of a lack of space or difficult angles. Because of its shape, the offset screwdriver can be used when there is very little clearance between the screw head and a surrounding surface. To redress the bits—there are two ends to the offset screwdriver—follow the same procedures as were described for the standard screwdriver.

Spiral Ratchet Screwdrivers

The *spiral ratchet screwdriver*, sometimes called a *Yankee drill* (see Figure 25-5), is

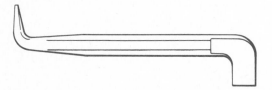

Figure 25-4. The offset screwdriver is used when space is limited.

Figure 25-5. A spiral ratchet screwdriver.

worked by pumping the handle up and down. The pumping action turns the bit. The spiral ratchet screwdriver is used to drive a screw into a pilot hole that has been drilled. Most spiral ratchet screwdrivers take several sizes of twist drills in addition to the screwdriver bit. These are redressed in the same manner as described in the section on twist drills.

Using and Caring For Pliers

Like screwdrivers, pliers are available in different designs and sizes for different kinds of jobs. Pliers are sized usually according to their length—5, 6, 8, and 10. The most common are the combination, diagonal cutting, chain-nose, needle-nose, duckbill, lock-grip, water pump, snapring, and lineman's pliers.

Combination Pliers

The combination pliers—sometimes called *slip-joint* or *side-cutting pliers*—has an adjustable jaw that opens to two positions and a wire cutter at the rear of the jaw. This common pliers is the one most often found—and used—in the shop, the farm, or the home. The combination pliers is used to hold things and to cut wire. It is not to be used as a wrench or screwdriver. If it is used for those purposes, it will inevitably damage the nut or the head of the bolt or screw.

Diagonal Cutting Pliers

The *diagonal cutting pliers*—sometimes called *dike's*—has a cutting jaw with a rounded back. It is used to cut wire or remove or replace cotter pins.

Chain-Nose, Needle-Nose, and Duckbill Pliers

The chain-nose, needle-nose, and duckbill pliers differ only in the shape of the jaw. The *needle-nose pliers* has a very thin jaw so that it can reach into very narrow and restricted areas. The *duckbill pliers* has a broad and rather flat jaw. It is used primarily to install washers, springs, and pins where vertical clearance is restricted. The *chain-nose pliers* has a jaw that is in between those of the other two. It is the most common.

Lock-Grip Pliers

The *lock-grip pliers*—or Vise Grip®—is designed to clamp metal together and to securely hold round stock. The lower jaw can be adjusted by turning a cap screw on the rear handle. Do not use a lock-grip pliers where the finish of the metal is important, because the pliers digs into the metal slightly for holding power.

Water Pump Pliers

The *water pump pliers* has a hinged jaw with a slip joint. The pliers can be adjusted to four or five different jaw sizes. It is excellent for holding pipe and round stock and for turning large packing nuts on pumps.

Snap-Ring Pliers

The *snap-ring pliers* has a small, pointed end that is at right angles to the handle of the pliers. A snap-ring allows the jaw to expand and contract inside an opening.

Lineman's Pliers

The *lineman's pliers* is used by electricians primarily to cut and form wire. It has both a heavy, flat jaw and a cutting jaw. The pliers is not intended to be used on wire that is carrying electric current. *Do not* use the pliers on current-carrying ("hot") wires, even though the pliers have an insulated covering on the handle.

Using and Caring For Wrenches

Four types of end wrenches are commonly used in agricultural industry: *adjustable open-end*, *open-end*, *box-end*, and *combination box and open-end*.

Adjustable Open-End Wrenches

The adjustable open-end wrench has a sliding lower jaw that is moved by an adjusting screw. The adjustable open-end wrench is a handy wrench because it can fit many different-sized nuts. The principal objection to using the adjustable open-end wrench is that it never fits any nut as snugly as a standard open-end wrench made for only one size nut.

The adjustable open-end wrench generally is available in lengths of 4 to 12 in [10.2 to 30.1 cm], although it is possible to get one as large as 24 in [60.1 cm]. The 12-in adjustable wrench is usually as big as needed on agricultural jobs. This size wrench has a capacity of $1^5/_{16}$ in [33.3 mm]. That is the width of the nut, *not* the diameter of the bolt.

The wrench opening is at a $22^1/_2$° angle to the wrench handle, so it is not very satisfactory in tight quarters.

There is a correct way to use the adjustable open-end wrench. Place the end of the wrench over the nut so that the pressure will be applied to the fixed jaw rather than the sliding jaw (see Figure 25-6). The fixed jaw can stand more force before it will break.

When the wrench opening is placed over the nut, tighten the adjusting screw while slightly rocking the handle. This will

TOOL REDRESSING AND CONDITIONING

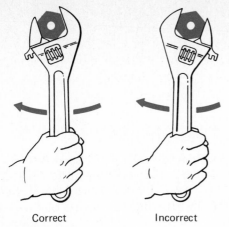

Correct   Incorrect

Figure 25-6. Always apply the pressure to the fixed jaw of the adjustable open-end wrench.

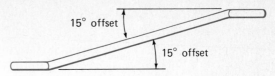

15° offset

15° offset

Figure 25-7. The offset wrench helps prevent injury to the mechanic's hands.

help ensure a snug fit. If the adjusting screw has a lock on it, secure the lock before beginning to turn the nut. This helps prevent the wrench from slipping and damaging the corner of the nut. Once a nut is rounded, it is almost impossible to turn.

Wrenches are heavy, but they are not designed to be used as hammers or crowbars. Using them in these ways will only damage them. Make sure the wrench is clean and that the jaws have no oil or grease on them. Occasionally lubricate the adjusting screw.

Standard Open-End Wrenches

The standard open-end wrench comes in a set of fixed sizes, from $1/4$- to $1^1/4$-in [6.4- to 31.2-mm] width of the opening. A complete set of wrenches generally consists of 20 wrenches. Each wrench usually has two open ends. One wrench, for example, may have an opening of $1/2$ in [12.7 mm] at one end and $9/16$ in [14.3 mm] at the other end.

On most wrenches, the opening is at a 15° angle to the handle. The 15° offset allows the worker to turn a hexagonal nut

when the swing of the wrench is limited to only 30°. That is about half of what normally would be possible if the end were not offset.

In tightening a nut with a right-hand thread and a limited swing, turn the nut clockwise as far as possible. Then turn the wrench over, and position the wrench on the nut as far counterclockwise as possible. Again begin to turn the nut in a clockwise direction. Of course, this procedure is reversed to loosen a nut. However, it is possible to get wrenches with openings at angles of $22^1/2°$, 30°, and even 60°.

It is always important to use the right-size wrench. A wrench that is too large will slip, damaging the nut and wasting a lot of motion. Do not use extensions, or "cheaters," to make a wrench longer, and don't try driving a wrench with a hammer.

Box-End Wrenches

A *box-end wrench* also is a fixed-size wrench, but the ends are closed (or boxed) instead of open (see Figure 25-8). The box-end wrench has three advantages over the other wrenches. First, the twelve-sided box fits very snugly over the nut. This

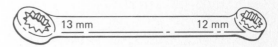

13 mm   12 mm

Figure 25-8. A box-end wrench helps prevent damage to the bolt or nut.

helps prevent damage to the nut. Second, because of the tight fit, the wrench is less likely to slip and cause workers to injure their hands. Finally, the box-end wrench can be used where there is only a 15° swing possible.

There is also a disadvantage. After each turn of the nut, it is necessary to lift the wrench completely off the nut before starting another turn. This doesn't sound like a problem, but when you have a lot of nuts to tighten or loosen, the extra time can be a problem.

The good mechanic knows the capacity for each wrench and does not abuse it by trying to overtorque (turn too tight), extend the wrench, or strike it with a hammer. Ignoring these three principles will break the wrench, ruin the nut, injure the worker, or do all three things.

A variation of the box-end wrench is the *tubing wrench* (see Figure 25-9). This wrench has a slot opening in the side of the end that allows the wrench to be fitted over a hydraulic line. The design allows the wrench to make very snug contact with fittings that are made of brass or other soft metals, reducing the possibility of damage. Notice in the figure that the force is applied against the full side of the wrench to eliminate spreading the opening.

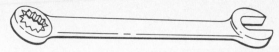

Figure 25-10. A combination wrench has both an open-end and a box-end.

Combination Wrenches

A combination box and open-end wrench, as might be expected, has one open end and one box end, both of the same size (see Figure 25-10). The advantage of the *combination wrench* is that a nut can be loosened with the box end and then quickly removed with the open end. The disadvantage is that both ends are of the same size, requiring the shop to stock more wrenches.

Socket Wrenches

Socket wrenches are available in a wide variety of sizes and capacities. There are three basic types of sockets—*hand*, *power*, and *impact*. Each differs in strength and is designed for a specific use. The hand socket should be used only on hand tools with limited torque. *Torque* is twisting pressure. The power socket is used with electric or air-powered (pneumatic) wrenches that have up to 250 ft · lb [34.6 m · kg] of torque. The impact socket is heavy and can transmit up to 1000 ft · lb [138.2 m · kg] of torque (see Figure 25-11).

Socket wrenches also are classified according to the number of points inside

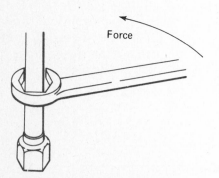

Figure 25-9. A tubing wrench is used on hydraulic couplings.

Hand   Power   Impact

Figure 25-11. Three common sockets.

TOOL REDRESSING AND CONDITIONING

6-point   8-point   12-point

Figure 25-12. Sockets are also classified by the number of points inside the socket.

Standard   Deep

Figure 25-13. Deep sockets are used to reach over long objects such as spark plugs.

the socket. There are 6-, 8-, and 12-point sockets (see Figure 25-12). The 6-point socket has a heavier wall and can be used on either power or impact wrenches. It is designed for six-sided (hexagonal) nuts. The 8-point socket is designed only for square nuts used on some farm machines. It will not fit hexagonal nuts. The 12-point socket is designed only for hand tools; it has a number of turning positions when space is limited. It should not be used on power or impact wrenches because of the thin side wall construction.

Another classification of sockets is according to the size of the drive. Each socket has a square opening in its back where the wrench fits. This opening is called the *drive*. Common drive sizes range from $\frac{1}{4}$ to 1 in. The drive size actually refers to the size of the square shaft which fits into the socket. For example, a $\frac{1}{4}$-in drive set has a $\frac{1}{4} \times \frac{1}{4}$ inch square drive and contains sockets to fit nuts from $\frac{3}{16}$ to $\frac{1}{2}$ in. A $\frac{3}{8}$-in drive set has sockets fitting nuts sized $\frac{3}{8}$ to $\frac{3}{4}$ in. The $\frac{1}{2}$-in drive set would fit nuts ranging in size from $\frac{7}{16}$ to $1\frac{1}{4}$ in. Special adapters are available to use $\frac{1}{4}$-in drive sockets on a $\frac{3}{8}$-in drive wrench or a $\frac{3}{8}$-in drive socket on a $\frac{1}{2}$-in drive wrench. In working on heavy agricultural equipment, it is desirable to have a $\frac{3}{4}$- or a 1-in drive set for larger sizes.

Another consideration in selecting sockets is that they may be sized according to inches or millimeters (metric system). A U.S. Customary System set of sockets may be sized from $\frac{3}{8}$ to 1 in. A similar set

sized according to the metric system may consist of sockets ranging from 9 to 19 mm.

Finally, sockets may be classed according to whether they are *standard* or *deep* (see Figure 25-13). Deep sockets have a much longer body and can reach over objects such as spark plugs and long studs. The special spark plug socket usually has a neoprene insert to help protect the plug from damage. A deep-socket set is essential for the mechanic working on vehicles or machinery powered by internal-combustion engines.

Socket Wrench Handles

Any socket wrench, of course, is of little value without a handle to make it work. There are generally four kinds of handles in the typical socket set: *ratchet*, *break-over* or *flex*, *sliding offset*, and *speed*. The handles are all sized according to drive size (see Figure 25-14).

Ratchet Handle. The ratchet is the most commonly used handle because of its ease of use. After you turn the nut to the end of the swing, the handle simply is "ratcheted" back to reengage the nut for another swing. The worker is not required to remove the socket from the nut between strokes. There are two kinds of ratchet handles—*standard* and *fine-tooth*. The standard ratchet has a swing of 15°. A fine-tooth ratchet can swing within a 4°

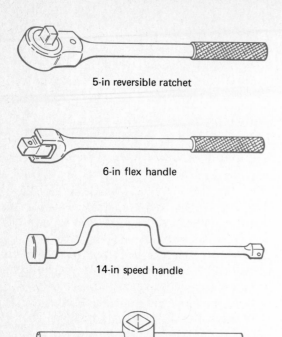

5-in reversible ratchet

6-in flex handle

14-in speed handle

Sliding offset handle

Figure 25-14.   Four kinds of socket wrench handles.

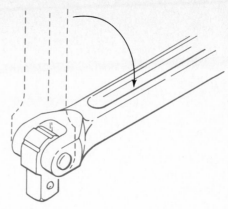

Figure 25-16.   A break-over or flex handle.

arc. This makes it very valuable in tight situations.

The direction of the rotation can be altered by changing the ratchet pawl—the sliding piece that locks the ratchet into position. Usually this is done by simply flip-

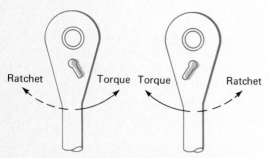

Ratchet   Torque Torque   Ratchet

Figure 25-15.   The direction of the rotation can be altered by changing the ratchet pawl.

ping a small lever on the back of the ratchet. The direction of the lever indicates the direction of the turn (see Figure 25-15).

Break-Over Handle.   The *break-over* or *flex handle* is a bar that can be shifted into different positions on a hinged joint. This enables the wrench to be used in different offset angles. Usually there is a pressure ball to stop the swing at 0° and 45° (see Figure 25-16). The break-over or flex handle is used primarily to loosen very tight nuts and stud bolts and for final tightening. A disadvantage of the handle is that the socket must be removed from the nut after each turn.

The sliding offset handle, or T handle, slips over a long, round bar (see Figure 25-17). A great deal of pressure can be applied this way, and a complete 360° swing is possible (although 180° is usually all that is accomplished in one turn). The sliding offset handle is used to loosen stubborn nuts and stud bolts. At the end of a swing, you can push the bar through the handle without removing the socket from the nut. An extension bar can be inserted into the handle to make more of a T handle. This allows even more pressure to be exerted.

TOOL REDRESSING AND CONDITIONING

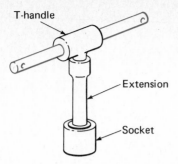

T-handle

Extension

Socket

Figure 25-17.   Using the sliding offset handle.

Speed Handle.   The speed handle is designed to tighten nuts quickly. It is shaped similarly to the carpenter's brace. The speed handle usually does not exert enough torque to make the nut as tight as it should be. It is used to run several nuts down on the bolts.

Two accessories that will add to the versatility of the socket wrench set are the *universal joint* and the *extension bar*. The universal joint allows the worker to get into places where a straight handle is almost impossible to use. Extension bars also make it possible to reach places where an ordinary socket wrench could not be used. The extension bars come in lengths from $1\frac{1}{2}$ to 20 in [3.8 to 51 cm].

Similar to the extension bar is the nut driver. It generally uses sockets of less than $\frac{1}{2}$ in [12.7 mm] and is good for working with a large number of small nuts.

Torque Wrenches

The *torque wrench* accurately measures the amount of torque, or twisting pressure, on a nut or stud bolt. The torque wrench is available with either conventional or metric scales. The conventional units measure in pound-feet (lb · ft) or pound-inches (lb · in). The metric units use meter-kilograms (m · kg) and centimeter-kilograms (cm · kg). The torque wrench is used when it is important to tighten nuts or bolts to exact specifications.

Three basic types of torque wrenches are available: *beam*, *dial*, and *click-type* (see Figure 25-18). The beam torque wrench relies on the worker driving the beam while taking a reading on the scale using a free-standing pointer. The torque may be read when you are tightening or loosening a nut or bolt. This type of torque wrench is generally the cheapest.

The dial torque wrench uses a direct-reading dial. While it is easy to read, it is also difficult to keep in adjustment. The

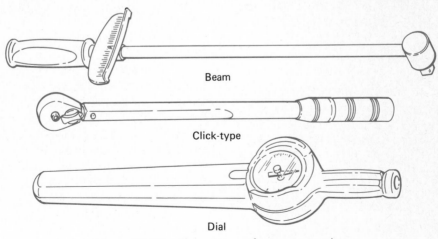

Beam

Click-type

Dial

Figure 25-18.   Three types of torque wrenches.

click-type torque wrench is preset to the specification desired for the job. When the torque is reached, the head snaps over with an audible click, and the worker knows the correct pressure has been applied. This wrench is favored by many mechanics because of its accuracy and speed.

The accuracy of the torque wrenches depends on the condition of the threads, how they have been lubricated, and how well they have been taken care of. These should be considered when you are torquing nuts.

Tables 25-1 and 25-2 show the grade markings for steel bolts and the torques for different sizes. This should be used as a guide in using a torque wrench.

Tool Care: A Review

As humans substitute machines for manual labor in agricultural industry, tools become an important factor in maintenance and repair.

The screwdriver is available as a standard, Phillips, clutch-head, offset, and spiral ratchet drive. Each is designed for a specific purpose. They should not be used as a pry bar or a chisel. When you use the screwdriver, select the proper size to fit the screw head. The bit should fit snugly in the slot. If redressing is necessary, restore the bit to the original thickness with a grinding wheel or a file.

Pliers are available in several sizes and designs for specific uses. The combination pliers is the most common; it is used for greater holding power and to cut wire. The slip-joint combination pliers may be adjusted to two sizes. Other common pliers include the diagonal cutting, chain-nose, needle-nose, duckbill, lock-grip, water pump, snap-ring, and lineman's pliers.

The four types of end wrenches are the adjustable open-end, open-end, box-end, and the combination box and open-end wrench. Each has a place in the maintenance and repair of equipment. Me-

TABLE 25-1. ASTM and SAE Grade Markings for Steel Bolts and Screws

GRADE MARKING	SPECIFICATION	MATERIAL
	ASTM—A 307 SAE—Grade 2	Low carbon steel Low carbon steel
	SAE—Grade 5 ASTM—A 449	Medium carbon steel, Quenched tempered
	SAE—Grade 7	Medium carbon alloy Steel, quenched tempered, roll-threaded after heat treatment
	SAE—Grade 8 ASTM—A 354 Grade BD	Medium carbon alloy Steel, quenched tempered

TOOL REDRESSING AND CONDITIONING

TABLE 25-2.   Maximum Torque Values for Four Grades of Steel Bolts

SIZE Threads per in.	SAE GRADE 2 TENSILE STRENGTH (74,000 min psi) TIGHTENING TORQUE lub. lb/ft	SAE GRADE 5 TENSILE STRENGTH (120,000 min psi) TIGHTENING TORQUE lub. lb/ft	SAE GRADE 7 TENSILE STRENGTH (133,000 min psi) TIGHTENING TORQUE lub. lb/ft	SAE GRADE 8 TENSILE STRENGTH (150,000 min psi) TIGHTENING TORQUE lub. lb/ft
1/4"				
20	4	6	8	9
28	5	7	9	10
5/16"				
18	8	13	16	18
24	9	14	18	20
3/8"				
16	15	23	30	35
24	17	25	35	40
7/16"				
14	24	35	45	55
20	25	40	50	60
1/2"				
13	35	55	70	80
20	40	65	80	90
9/16"				
12	55	80	100	110
18	60	90	110	130
5/8"				
11	75	110	140	170
18	85	130	160	180
3/4"				
10	130	200	240	280
16	145	220	280	320
7/8"				
9	125	320	400	460
14	140	350	440	500
1"				
8	190	480	600	680
12	200	530	660	740

chanics prefer the box-end wrench because it fits tightly and they can work in a limited space.

Socket wrenches have the advantages of the box-end wrench plus the speed of the ratchet. The three sockets commonly used in agricultural industry are the hand, power, and impact sockets. They are also sized according to the size of the drive and the depth of the socket. The socket may be driven with a ratchet, break-over, sliding offset, or a speed handle.

When it is important to tighten nuts or bolts to exact specifications, a torque wrench should be used. The torque wrench is available as a beam, dial, or a break-away design. Most engine-repair manuals specify the tightness of most nuts and bolts. By following these specifications, exact tolerances may be established.

THINKING IT THROUGH

1. List the five types of screwdrivers.
2. What are the common types of pliers used in agricultural industry?
3. List the four types of end wrenches and describe where each type should be used.
4. What type of socket should be used on a square nut?
5. What are the advantages of a 12-point socket over a 6-point socket?
6. What are the advantages of a fine-tooth ratchet wrench? What are the disadvantages?
7. When should a torque wrench be used?
8. What are the three types of torque wrench commonly used in agricultural industry?

GLOSSARY

Acetone A fragrant liquid used in cylinders to dissolve acetylene gas.

Acetylene A volatile (unstable) gas composed of carbon and hydrogen and used to produce high temperatures for welding and cutting.

Agricultural mechanics The operation and maintenance of the machines and equipment used in agricultural industry.

Alloy A product made by the combination of two or more metals.

Amperage The measure of the electrical current (in amperes) flowing through a circuit.

Anneal To heat metal until it is softened and relieved of stress.

Architects' scale A triangular 12-inch rule with divisions in feet and inches used to prepare scale drawings.

Arc welding Fusion welding using electrical resistance to heat and melt the metals.

Assembly drawings A set of drawings that show how to put together and take apart equipment or other objects.

Backflash A hissing sound that results from a restriction in the welding or cutting torch.

Base metal The pieces of metal composing the parts to be welded.

Bastard file A file with a rough or coarse cut.

Bead The mass that forms as the puddle is moved along by an operator during welding.

Bevel Angle, as of a cut.

Blueprint A set of architectural plans or technical drawings; a copy of an original drawing using a special nonduplicating process.

Board A piece of wood less than 2 inches thick.

Board foot A unit of measure that is 1 inch thick and 1 foot wide and 1 foot long.

Brass A yellow metal alloy of copper combined with 10 to 40 percent zinc and small amounts of other metals.

Brazing Welding with an alloy which melts above 800°F (427°C) and below the melting temperature of the base metal.

Break line A line used to indicate that part of a drawing has been omitted.

Bronze A reddish-yellow metal alloy of copper combined with tin and small amounts zinc.

Carburizing or reducing flame A flame with an excess of acetylene used to heat metal, soft-solder, and hard-face.

Cast iron Iron containing 1.7 percent or more carbon which is then poured into a mold.

Combination blade A saw blade used for both ripping and crosscutting lumber.

Commercial lumber Wood that has been dressed and graded by a large saw mill.

Coniferous trees Trees that do not shed their leaves at one time, such as pine and cedar.

Conventions Symbols commonly used in a drawing.

Countersink To set the head of a screw even with or slightly below the wood surface.

Crack To open a cylinder valve quickly before use to blow out any dirt, foreign material, or to check for excessive acetone.

Crosscutting To cut across the grain of wood.

Cutting bill A list of each item (with the exact length) needed for a plan.

Cylinder Container used to store gasses, such as acetylene or oxygen.

Dado A groove that is cut or chiseled across a board to form a joint.

Deciduous trees Trees that shed their leaves during a specific season, such as oak and maple.

Detail drawing A drawing of a small part of a larger project; usually drawn to a different (larger) scale than that of the whole project.

Dimension line A line on a scale drawing showing the measurements between points on an object.

Dimension lumber Lumber between 2 and 5 inches thick.

Drawfiling Filing at a right angle to the metal.

Drive The opening in a socket where a wrench fits.

Ductile Able to be formed or worked easily.

Electrode A solid conductor which provides filler metal and through which the arc welding current flows.

Engineers' scale A triangular 12-inch rule with divisions in feet and decimal parts of a foot used to prepare scale drawings.

Ferrous metals   Metals which contain iron.

File   A small, hand-held tool used to shape and smooth wood or metal.

File card   A brush used to clean a file.

Flashback   A hazzard occuring when the fuel pressure is too low resulting in the oxy-fuel flame burning back into the torch and hose.

Flux   A coating used on the electrode to clean the base metal and provide a gaseous shield.

Forestry   The science dealing with planting, managing, and harvesting trees.

Full-scale   Exactly proportional and to an object's actual size, as in a full-scale drawing.

Gain   A notch cut into one piece of wood so another piece can be fastened to it.

Galvanize   To coat with zinc to prevent rusting.

Gas welding   Joining metals either by fusion or cohesion; the gases used are acetylene and oxygen.

Grain   The pattern of the alternating layers of spring and summer wood.

Grid paper   Paper with lines crossing each other throughout, used for drawing to scale.

Grit   An indication of the degree of coarseness of a grinding wheel or stone.

Hard-face or hard-surface   To add a thin layer of a metal alloy on a base metal to protect from impact or abrasion.

Hard-solder   Braze welding.

Hardwood   Wood that comes from deciduous trees.

Horticulture   The science involved with the production and merchandising of fruits, vegetables, flowers, nursery stock, and turf grass.

Hose   Flexible tubing used to carry gas from the cylinder to the welding torch.

Interior lines   Lines added to finish a rough drawing.

Job cluster   A group of related jobs which use many of the same competencies.

Kerf   The cut made by a saw.

Leader line   A line which refers (leads) to a particular point on a drawing.

Legs   The extensions of a fillet weld.

Lumber   Dressed or processed wood ready for use in building.

Machine   To shape or finish to a specific shape with the aid of mechanical tools.

Malleable   Able to be easily bent, shaped, welded, drilled, or sawed.

Metric scale   A triangular rule 30 cm long used to show metric proportions.

Metric system   A decimal system of measurements using the meter and the kilogram as base units.

Mild steel   Steel that contains between 0.25 and 0.60 percent carbon.

Neutral flame   A flame used in gas welding consisting of equal parts of oxygen and acetylene.

Nonferrous metals   Metals containing little or no iron.

Order bill   A list of materials needed for a plan and given in standard supplier's measurements.

Orthographic drawing   A drawing made at right angles to the object showing all sides and views.

Oxidize   To combine with oxygen; to burn off impurities in iron by heating to a very high temperature.

Oxidizing flame   A flame with an excess of oxygen used in special welding applications.

Oxygen   An odorless, colorless, tasteless gas used to support combustion.

Pallet   A small platform used for moving goods.

Parts list   A list of parts shown in an assembly drawing.

Penetration   The depth of fusion into a base metal expressed as a percentage of the thickness of the base metal.

Pig iron   Iron directly from the blast furnace which is cast in blocks or "pigs."

Plan   A drawing or sketch communicating ideas or instructions to someone else.

Ply   The grain or layer of a sheet of material.

Plywood   A building material made of alternating layers of wood glued together.

Polarity   The direction of current flow in an electrical circuit.

Preheating   Heat applied before welding, cutting, or forming is performed.

Primer coat   A coat of paint to seal and bond surfaces.

Puddle A small area of liquid metal controlled during the welding process.

Quench To quickly cool metal by putting it into a liquid.

Rabbet A channel or groove along the edge or end of a piece of wood.

Redress To maintain a cutting tool; to sharpen or recondition a tool.

Regulator A device that reduces the pressure of gas from a cylinder.

Resistance The opposition to the current flow in an electrical circuit.

Rip To cut a board lengthwise or parallel to the grain.

Scale The proportion of the size of the drawing to the dimensions of the finished object.

Scale drawing A drawing in proportion; for example, on a scale drawing where 1″ = 1 foot, every inch represents 1 foot on the object.

Schematic drawing A diagram showing the relationship of one object to another.

Sear To quickly heat the surface of cast iron to reduce the carbon film.

Sketch A hand drawing or visual plan.

Skidder A machine used to load logs for transporting.

Slag A protective coating formed over the liquid metal during arc welding.

Soft-solder To join metals at less than 800°F (427°C).

Softwood Wood that comes from coniferous trees.

Solder An alloy made of lead and tin and small amounts of other metals; to join metals using a soft-solder process.

Stainless steel A special steel containing chromium, nickel, and other metals which resists rusting.

Steel Iron which has been refined and combined with a desirable amount of carbon; to sharpen a knife or cutting tool.

Strop To finish sharpening a knife by sweeping the blade across a leather piece.

Tap A tool used to cut an internal thread.

Temper To toughen metal by heating and slowly cooling.

Thread Machining a thread on a screw, nut, or bolt.

Timber Wood at least 5 inches thick and wide.

Tinning A process of plating the surface of a joint before joining.

Title block A small section, usually in the lower right-hand corner of a drawing, which contains information for reading the drawing or using the plan.

Tool steel Steel that contains between 0.5 to 1.5 percent carbon and can be tempered and hardened.

Transfer lettering Printed letters that can be applied to a surface by rubbing the backing to which the letters are attached.

Turn-key job A complete project from start to finish, such as the building of a milking parlor in which the planning, building, and installation of equipment is done by an outside company.

Ungraded lumber Wood that has been sawed at a small mill; it is generally not dressed and graded.

Upset To reduce the length of metal and increase its diameter.

Unstable gas A gas that decomposes or breaks down easily.

U.S. Customary System The system of measurement currently used in the United States consisting of pounds and feet as the base units. (*See* Metric.)

Whet To sharpen to a fine edge.

Wrought iron Almost pure iron that is very malleable.

INDEX